Non-equilibrium Hydromagnetic Dynamos

Online at: https://doi.org/10.1088/978-0-7503-6362-4

Non-equilibrium Hydromagnetic Dynamos

Krzysztof A Mizerski
Polish Academy of Sciences, Institute of Geophysics, Warsaw, Poland

IOP Publishing, Bristol, UK

ISBN 978-0-7503-6362-4 (ebook)
ISBN 978-0-7503-6358-7 (print)
ISBN 978-0-7503-6359-4 (myPrint)
ISBN 978-0-7503-6361-7 (mobi)

DOI 10.1088/978-0-7503-6362-4

Version: 20231201

IOP ebooks

British Library Cataloguing-in-Publication Data: A catalogue record for this book is available from the British Library.

Published by IOP Publishing, wholly owned by The Institute of Physics, London

IOP Publishing, No.2 The Distillery, Glassfields, Avon Street, Bristol, BS2 0GR, UK

US Office: IOP Publishing, Inc., 190 North Independence Mall West, Suite 601, Philadelphia, PA 19106, USA

To my dearest

Agata

Contents

Preface

The effect of a hydromagnetic dynamo is very common in nature. The majority of astrophysical magnetic fields, including the terrestrial field, are generated by turbulent flow of a conducting liquid, such as plasma or liquid metal in the interiors of planets, stars, accretion discs, etc The flow is typically driven thermally, compositionally (cf e.g. [1–4]) or by development of violent instabilities such as, e.g. the magneto-rotational instability (cf [5, 6]). If conditions for dynamo action at large scales are satisfied, that is, in particular, the fluid domain is not too small and the turbulent flow is vigorous and lacks reflectional symmetry, the energy of the flow is then partially transformed into the energy of large-scale magnetic fields. The large-scale field amplification and saturation rely on formation of large-scale electromotive force (EMF) in the system, however, there are various mechanisms that can lead to generative mean EMF, i.e. capable of generation of large-scale fields. The $\mathbf{\Omega}(\mathbf{x}) \times \mathbf{J}$ effect with non-uniform background rotation $\mathbf{\Omega}(\mathbf{x})$ (cf [7–11]), the 'shear-current' effect (cf [12–14]) or the cross-helicity dynamo (cf [15–18]) rely on inhomogeneity of the large-scale flows.

However, the turbulent mechanism of dynamo action, which is typically invoked to be responsible for generation of large-scale fields is the so-called α-effect, based on nonlinear interactions of the small-scale fluctuating components of the turbulent state, which generate the large-scale EMF (cf e.g. [19–25]). The latter can also be combined with the effect of non-uniform background rotation producing then the dynamo of $\alpha\mathbf{\Omega}$ type; the shear which strains and advects the magnetic field lines in the $\mathbf{\Omega}$ effect is, however, often unnecessary and dynamos of α^2 type are known. For an α-type dynamo action to occur, the wave field must exhibit chirality (lack of reflexional symmetry), and this requires some mechanism (e.g. gravity $\mathbf{g}$ and density stratification in conjunction with mean rotation $\mathbf{\Omega}$, or boundary forcing) that breaks the 'up-down' symmetry of the field and preferentially excites waves with one helicity sign; a useful measure of the flow chirality is the kinetic helicity, $\langle \mathbf{U} \cdot \nabla \times \mathbf{U} \rangle$, where $\langle \cdot \rangle$ denotes the ensemble mean and another pseudoscalar quantity of importance in dynamo theory is the cross-helicity $\langle \mathbf{U} \cdot \mathbf{B} \rangle$, cf e.g. [17, 26] and [18] for a review. An electromotive force $\mathcal{E}_i = \alpha_{ij} \langle B \rangle_j$ is then generated, where $\langle \mathbf{B} \rangle$ is the mean magnetic field, i.e. an α-effect in the terminology of Steenbeck *et al* (1966) [22], and this leads to exponential growth of magnetic energy until the growing Lorentz force reacts back upon the wave field, leading to a saturated state ([27]). In the standard approaches this mechanism relies on a phase shift between the kinetic and magnetic components of turbulent fluctuations, which is generated by the fluid's resistivity. In other words such a mechanism becomes ineffective in low-resistivity fluids/plasma. It has been demonstrated in [27] and chapter 12 of Moffatt and Dormy (2019) [25] and thus known for a long time, that an oscillatory forcing with just one frequency of oscillation induces a non-decaying wave in a resistive fluid, which can create a mean EMF through self-interactions between the kinetic and magnetic components of a single wave; the resistivity is responsible for creation of a necessary phase shift between the small-scale velocity and small-scale magnetic field of the wave. The motivation for the study presented in this book was based on the simple fact, that every force can be decomposed into spatial and temporal Fourier modes, however, such an expansion involves many waves with different frequencies of oscillation and thus interactions between them are allowed.

The investigations of the large-scale dynamo mechanisms in magnetohydrodynamic (MHD) turbulence, that is, those that lead to generation of large-scale magnetic fields, has been mainly limited to equilibrium, i.e. statistically stationary turbulence, which excludes interactions between waves with distinct frequencies. However, it is important to realize the significant role of non-equilibrium effects in turbulent dynamos since in real systems turbulence is typically not in equilibrium, i.e. couplings between distinct waves occur and mean quantities such as the mean energy and helicity are typically time-dependent. In a sequence of recent papers [28–32], a new dynamo mechanism has been described (originally identified in [33]), based on interactions between waves induced by a forcing consisting of more than just one frequency of oscillation, which operates in the dynamical regime, thus under inclusion of the effect of the Lorentz force. The new dynamo mechanism is described as 'fast' in the terminology of Vainshtein and Zel'dovich (1972) [34], that is, it leads to magnetic field amplification at a rate independent of resistivity, when the resistivity tends to zero. The EMF survives in the limiting situation $\eta = 0$, leading to an α-effect even in ideal fluids, whereas in a low-resistivity plasma the new mechanism can produce a dominant contribution to the mean EMF. It is supposed, therefore, that the presented theory might be relevant to the dynamo generation of magnetic fields in extremely highly conducting plasma, such as in the galactic interstellar medium or in galaxy clusters, in particular, in the early Universe, when the galaxies were formed. Indeed, estimates of molecular resistivity for the Milky Way Galaxy, $\eta_{\text{galaxy}} \approx 10^3\ \text{m}^2\ \text{s}^{-1}$ and galaxy clusters $\eta_{\text{cluster}} \approx 3 \times 10^{-3}\ \text{m}^2\ \text{s}^{-1}$, taken from [35] and [36], respectively, lead to extremely long characteristic diffusion times for the magnetic field on the entire length scale of the system, $t_{\text{galaxy}} \approx 10^{30}$ y, and $t_{\text{cluster}} \approx 10^{34}$–$10^{38}$ y, many orders of magnitude longer than the age of the Universe $\sim 10^{10}$ y. It seems therefore legitimate to neglect the effect of resistivity in such systems and study the generation of large-scale magnetic fields in ideal plasma (a 'fast dynamo' in the terminology of Vainshtein and Zel'dovich 1972 [34]), however, the new mechanism is also studied in the presence of weak resistivity. Furthermore, it is worth mentioning that according to the so-called primordial theory of the origin of galactic magnetic fields, the main build-up of the large-scale fields in such systems had taken place billions of years ago, when the galaxies were formed and it is only maintained by the turbulent flow in the later evolution; the dynamo no longer needs to operate because the plasma is nearly perfectly conducting and the field cannot decay during the Galaxy/cluster lifetime, because of the extremely low resistivity (cf [37]). Such a field was termed a 'fossil field' in [38], where the effect of weak plasma viscosity was investigated. In particular, a force-free mode of the primordial field should survive, as there are no mechanisms of magnetic energy decay, which could suppress such a mode. It should be emphasized, however, that the effect of gravity, which at the early stage of evolution of the galactic magnetic fields was significant, since it led to strengthening the magnetic field by bringing the field lines closer together, is neglected here for clarity of presentation of the new studied mechanism.

Another particularly interesting possibility for application of the developed theory is in engineering. Recent advancement of tokamak devices, allowed for reaching very high plasma temperatures, thus giving hope for the production of thermonuclear fusion power (cf [39, 40]). The effect of 'beating waves', i.e. interactions of waves with distinct but close frequencies explained in this book may prove to be useful for increasing the magnitude of the dynamo-induced magnetic field in tokamaks and thus the temperature and perhaps even the level of 'plasma entrapment', hence its stability.

This book reviews and synthesizes the recent findings of [28–32] concerning the dynamo effect of beating waves and yet another non-stationary dynamo effect associated with history of evolution of turbulent kinetic and cross helicities identified in [41]. The latter is based on coexistence of the kinetic and cross helicities in plasma turbulence and therefore may also be a promising candidate for enhancement of the dynamo-induced magnetic field in tokamaks, where due to the geometry of the flow and magnetic fields the system should be typically abundant in both aforementioned helicities. The most important aim of this book is to explain and emphasize the versatile role of non-equilibrium effects in turbulent hydromagnetic dynamos, hitherto typically neglected. On a final note we recall that the problem of large-scale EMF generation by random Lehnert waves [42] (also called the MC (Magneto-Coriolis) waves after [43]) has been addressed previously, e.g. by [44, 45] and including density stratification by [46]. However, the latter was a kinematic study and the previous two considered situations corresponding to the assumption of stationarity of turbulence, therefore no interactions between waves with distinct frequencies have been studied. Interestingly, a trace of the effect of non-stationarity of turbulence on the mean EMF in dynamo theory can be found in [16] (section 9.6.2, equation (9.232) of that book); however, it is ignored in the following analysis and stationarity is invoked.

We start with derivation of the magnetohydrodynamic equations and then in chapter 2 with a concise introduction to mean-field dynamo theory, with particular attention to the fully nonlinear turbulent dynamo process and the effect of strong turbulence on the form of the large-scale EMF. The dependence of the mean EMF on the turbulent kinetic, current and cross helicities is demonstrated and their production laws in magnetohydrodynamic turbulence are derived.

References

[1] Roberts P H and Soward A M 1972 Magnetohydrodynamics of the Earth's core *Ann. Rev. Fluid Mech.* **4** 117–54

[2] Brandenburg A and Subramanian K 2005 Astrophysical magnetic fields and nonlinear dynamo theory *Phys. Rep.* **417** 1–209

[3] Dormy E and Soward A M 2007 *Mathematical Aspects of Natural Dynamos* (Boca Raton, FL: Chapman and Hall)

[4] Roberts P H and King E M 2013 On the genesis of the Earth's magnetism *Rep. Prog. Phys.* **76** 096801

[5] Balbus S A and Hawley J F 1991 A powerful local shear instability in weakly magnetized disks. I–Linear analysis *Astrophys. J.* **376** 214–22
[6] Balbus S A and Hawley J F 1991 A powerful local shear instability in weakly magnetized disks. II—nonlinear evolution *Astrophys. J.* **376** 223–33
[7] Rädler K-H 1969 A new turbulent dynamo. I *Monats. Dt. Akad. Wiss.* **11** 272–9 [English translation: Roberts P H and Stix M 1971 The turbulent dynamo: a translation of a series of papers by F Krause, K-H Rädler and M Steenbeck *NCAR Technical Note NCAR/TN-60 +IA* pp 301–8]
[8] Roberts P H 1972 Kinematic dynamo models *Phil. Trans. R. Soc.* A **272** 663–98
[9] Moffatt H K and Proctor M R E 1982 The role of the helicity spectrum function in turbulent dynamo theory *Geophys. Astrophys. Fluid Dyn.* **21** 265–83
[10] Rädler K-H 1986 Investigations of spherical mean-field dynamo models *Astron. Nachr.* **307** 89–113
[11] Rädler K-H, Kleeorin N and Rogachevskii I 2003 The mean electromotive force for MHD turbulence: the case of a weak mean magnetic field and slow rotation *Geophys. Astrophys. Fluid Dyn.* **97** 249–68
[12] Rogachevskii I and Kleeorin N 2003 Electromotive force and large-scale magnetic dynamo in a turbulent flow with a mean shear *Phys. Rev.* E **68** 036301
[13] Rogachevskii I and Kleeorin N 2004 Nonlinear theory of a 'shear-current' effect and mean-field magnetic dynamos *Phys. Rev.* E **70** 046310
[14] Rogachevskii I, Kleeorin N and Liverts E 2006 Nonlinear shear-current dynamo and magnetic helicity transport in sheared turbulence *Geophys. Astrophys. Fluid Dyn.* **100** 537–57
[15] Yoshizawa A 1990 Self-consistent turbulent dynamo modeling of reversed field pinches and planetary magnetic fields *Phys. Fluids* B **2** 1589–600
[16] Yoshizawa A 1998 *Hydrodynamic and Magnetohydrodynamic Turbulent Flows: Modelling and Statistical Theory* (Dordrecht: Springer)
[17] Yokoi N 2013 Cross helicity and related dynamo *Geophys. Astrophys. Fluid Dyn.* **107** 114–84
[18] Yokoi N 2023 Unappreciated cross-helicity effects in plasma physics: anti-diffusion effects in dynamo and momentum transport *Rev. Mod. Plasma Phys.* (submitted, arXiv:2303.01834)
[19] Parker E N 1955 Hydromagnetic dynamo models *Astrophys. J.* **122** 293–314
[20] Braginskii S I 1964 Self-excitation of a magnetic field during the motion of a highly conducting fluid *Sov. Phys. JETP* **20** 726–35
[21] Braginskii S I 1964 Theory of the hydromagnetic dynamo *Sov. Phys. JETP* **20** 1462–71
[22] Steenbeck M, Krause F and Rädler K-H 1966 A calculation of the mean electromotive force in an electrically conducting fluid in turbulent motion, under the influence of Coriolis forces *Z. Naturforsch.* **21** 369–76 [English translation: Roberts & Stix (1971) pp 29–47]
[23] Krause F and Rädler K-H 1980 *Mean-field Magnetohydrodynamics and Dynamo Theory* (Oxford: Pergamon)
[24] Roberts P H 1994 Fundamentals of dynamo theory *Lectures on Solar and Planetary Dynamos (Publications of the Newton Institute)* ed M Proctor and A Gilbert (Cambridge: Cambridge University Press) pp 1–58
[25] Moffatt H K and Dormy E 2019 *Self-exciting Fluid Dynamos* (Cambridge: Cambridge University Press)
[26] Hamba F and Tsuchiya M 2010 Cross-helicity dynamo effect in magnetohydrodynamic turbulent channel flow *Phys. Plasmas* **17** 012301

[27] Moffatt H K 1972 An approach to a dynamic theory of dynamo action in a rotating conducting fluid *J. Fluid Mech.* **53** 385–99
[28] Mizerski K A 2018 Large-scale hydromagnetic dynamo by Lehnert waves in nonresistive plasma *SIAM J. Appl. Math. (SIAP)* **78** 1402–21
[29] Mizerski K A 2018 Large-scale dynamo action driven by forced beating waves in a highly conducting plasma *J. Plasma Phys.* **84** 735840405
[30] Mizerski K A 2020 Renormalization group analysis of the turbulent hydromagnetic dynamo: effect of non-stationarity *Astrophys. J. Suppl. Ser.* **251** 21
[31] Mizerski K A 2021 Possible role of non-stationarity of magnetohydrodynamic turbulence in understanding of geomagnetic excursions *Symmetry* **13** 1881
[32] Mizerski K A 2022 Dynamo effect caused by non-stationary turbulence in strongly magnetized, hot, low-density plasma *Astron. Astrophys.* **660** A110
[33] Mizerski K A, Bajer K and Moffatt H K 2012 The mean electromotive force generated by elliptic instability *J. Fluid Mech.* **707** 111–28
[34] Vainshtein S I and Zel'dovich Y B 1972 Origin of magnetic fields in astrophysics *Sov. Phys. - Usp.* **15** 159–72
[35] Spitzer L 1962 *Physics of Fully Ionized Gases* (New York: Wiley)
[36] Schekochihin A A and Cowley S C 2006 Turbulence, magnetic fields, and plasma physics in clusters of galaxies *Phys. Plasmas* **13** 056501
[37] Zweibel E G and Heiles C 1997 Magnetic fields in galaxies and beyond *Nature* **385** 131–6
[38] Mizerski K A and Moffatt H K 2018 Dynamo generation of a magnetic field by decaying Lehnert waves in a highly conducting plasma *Geophys. Astrophys. Fluid Dyn.* **112** 165–74
[39] Li J, Ni M and Lu Y 2019 The frontier and perspective for tokamak development *Nat. Sci. Rev.* **6** 382–3
[40] Gibney E 2022 Nuclear-fusion reactor smashes energy record *Nature* **602** 371
[41] Mizerski K A, Yokoi N and Brandenburg A 2023 Cross-helicity effect on α-type dynamo in non-equilibrium turbulence *J. Plasma Phys.* **89** 905890412
[42] Lehnert B 1954 Magnetohydrodynamic waves under the action of the Coriolis force *Astrophys. J.* **119** 647–54
[43] Finlay C C 2008 Waves in the presence of magnetic fields, rotation and convection *Les Houches, Dynamos* ed P Cardin and L F Cugliandolo (Amsterdam: Elsevier) vol 88 pp 403–50
[44] Moffatt H K 1970 Dynamo action associated with random inertial waves in a rotating conducting fluid *J. Fluid Mech.* **44** 705–19
[45] Soward A M 1975 Random waves and dynamo action *J. Fluid Mech.* **69** 145–77
[46] Walder M, Deinzer W and Stix M 1980 Dynamo action associated with random waves in a rotating stratified fluid *J. Fluid Mech.* **96** 207–22

Acknowledgements

This work was supported by a subsidy from the Polish Ministry of Education and Science for the Institute of Geophysics, Polish Academy of Sciences and by the Polish National Science Centre (Grant No. 2017/26/E/ST3/00554).

Author biography

Krzysztof Mizerski

Krzysztof Mizerski is an Associate Professor at the Department of Magnetism, Institute of Geophysics, Polish Academy of Sciences, ul. Ksiecia Janusza 64, 01-452, Warsaw, Poland. He has experience in theoretical fluid mechanics, in particular, the theory of the hydromagnetic dynamo and stratified convection. All of his works have been concerned with analytic approaches to hydrodynamic and magnetohydrodynamic problems supplied by numerical models. These involved investigations of singular structures in boundary layers, linear and nonlinear stability theory and description of fully developed, turbulent stratified convection. One of the main topics of the research conducted by Krzysztof Mizerski is the description of large-scale hydromagnetic dynamos in the fully nonlinear regime, in particular via the renormalization group method. His most recent works were dedicated to creation of a consistent dynamical picture of the hydromagnetic dynamo process induced by non-equilibrium effects in magnetohydrodynamic turbulence. He is an author of the textbook *Foundations of Convection with Density Stratification*, 2021, Springer, Cham.

Symbols

$\hat{\mathbf{e}}_i$	Cartesian unit vector
$\mathbf{U}$	velocity field, [m s^{-1}]
$\mathbf{B}$	magnetic field normalized by $\sqrt{\mu_0\rho}$, [m s^{-1}]
$\mathbf{W}$	vorticity field, $\mathbf{W} = \nabla \times \mathbf{U}$, [s^{-1}]
$\mathbf{J}$	electric current field normalized by $\sqrt{\rho/\mu_0}$, $\mathbf{J} = \nabla \times \mathbf{B}$, [s^{-1}]
$\mathbf{A}$	magnetic vector potential, $\mathbf{B} = \nabla \times \mathbf{A}$, [m^2 s^{-1}]
$\mathbf{E}$	electric field normalized by $\sqrt{\mu_0\rho}$, [m^2 s^{-2}]
p	pressure field, [N m^{-2}]
$\langle\mathbf{U}\rangle$	mean velocity field, [m s^{-1}]
$\langle\mathbf{B}\rangle$	mean magnetic field normalized by $\sqrt{\mu_0\rho}$, [m s^{-1}]
$\langle\mathbf{W}\rangle$	mean vorticity field, $\langle\mathbf{W}\rangle = \nabla \times \langle\mathbf{U}\rangle$, [s^{-1}]
$\langle\mathbf{J}\rangle$	mean electric current field normalized by $\sqrt{\rho/\mu_0}$, $\langle\mathbf{J}\rangle = \nabla \times \langle\mathbf{B}\rangle$, [s^{-1}]
$\langle\mathbf{A}\rangle$	mean magnetic potential, $\langle\mathbf{B}\rangle = \nabla \times \langle\mathbf{A}\rangle$, [m^2 s^{-1}]
$\langle p\rangle$	mean pressure field, [N m^{-2}]
$\mathbf{u}'$	turbulent velocity fluctuation, [m s^{-1}]
$\mathbf{b}'$	turbulent magnetic fluctuation normalized by $\sqrt{\mu_0\rho}$, [m s^{-1}]
$\mathbf{w}'$	turbulent vorticity fluctuation, $\mathbf{w}' = \nabla \times \mathbf{u}'$, [s^{-1}]
$\mathbf{j}'$	turbulent electric current fluctuation normalized by $\sqrt{\rho/\mu_0}$, $\mathbf{j}' = \nabla \times \mathbf{b}'$, [s^{-1}]
$\mathbf{a}'$	turbulent magnetic potential fluctuation, $\mathbf{b}' = \nabla \times \mathbf{a}'$, [m^{-2} s^{-1}]
$\mathbf{a}'_{\text{tor}}$	toroidal magnetic potential fluctuation, [m^{-2} s^{-1}]
$\mathbf{a}'_{\text{pol}}$	poloidal magnetic potential fluctuation, [m^{-2} s^{-1}]
$\mathbf{e}'$	turbulent electric field fluctuation normalized by $\sqrt{\mu_0\rho}$, [m^2 s^{-2}]
p'	turbulent pressure field fluctuation, [N m^{-2}]
$\mathbf{f}$	stirring force, [m s^{-2}]
D_0	forcing energy parameter defined in (3.132), [m^2 s^{-3}]
D_1	forcing helicity parameter defined in (3.132), [m s^{-3}]
Δ	'non-stationarity' function defined in (3.86) and (3.132), [s]
Γ	'non-stationarity' parameter defined in (3.86) and (3.132), [*nondimensional*]
$\tilde{\omega}$	'non-stationarity' parameter corresponding to beating frequency of interacting waves defined in (3.86) and (3.132), [s^{-1}]
$\mathbf{\Omega}$	background rotation vector, [s^{-1}]
σ_e	uniform electrical conductivity of the fluid, [A Vm^{-1}]
μ_0	magnetic permeability of vacuum, $\approx 1.26 \times 10^{-6}$ [N A^{-2}]
ρ	uniform density of the fluid, [kg m^{-3}]
ν	uniform kinematic viscosity of the fluid, [m^{-2} s^{-1}]
η	uniform magnetic diffusivity of the fluid $\eta = 1/\mu_0\sigma_e$, [m^{-2} s^{-1}]

τ_t	turn-over time of most energetic eddies, [s]
$\langle\cdot\rangle$	ensemble mean
$\langle\cdot\rangle_{\mathbf{x}}$	spatial average defined in (2.64)
$\langle\cdot\rangle_t$	temporal average defined in (6.4)
$\langle\cdot\rangle_c$	conditional subensemble average defined in (4.54a–c)
$\frac{\mathrm{D}}{\mathrm{D}t}$	total Lagrangian derivative $\frac{\partial}{\partial t} + \mathbf{U}\cdot\nabla$
$\frac{D}{Dt}$	mean Lagrangian derivative $\frac{\partial}{\partial t} + \langle\mathbf{U}\rangle\cdot\nabla$
$\frac{\hat{D}}{DT}$	co-rotational derivative defined in (5.12)
G_{ij}	mean velocity gradient $\frac{\partial\langle U\rangle_i}{\partial x_j}$, [s^{-1}]
Γ_{ij}	mean magnetic-field gradient $\frac{\partial\langle B\rangle_i}{\partial x_j}$, [s^{-1}]
L	size of the entire system, [m]
ℓ_e	size of the most energetic turbulent eddies, [m]
ℓ_ν	viscous length-scale cut-off for turbulent kinetic energy, [m]
ℓ_η	resistive length-scale cut-off for turbulent magnetic energy, [m]
ℓ	fluctuational length scales $\ell_e \leqslant \ell \leqslant \ell_\nu, \ell_\eta$, [m]
$\hat{f}$	Fourier transform of f defined in (2.14a–d)
$\mathbf{k}$	wavevector of fluctuations $k = \frac{2\pi}{\ell}$, [m^{-1}]
K_e	wavenumber of most energetic turbulent eddies $K_e = \frac{2\pi}{\ell_e}$, [m^{-1}]
K_ν	viscous wavenumber cut-off for spectral kinetic energy $K_\nu = \frac{2\pi}{\ell_\nu}$, [m^{-1}]
K_η	resistive wavenumber cut-off for spectral magnetic energy $K_\eta = \frac{2\pi}{\ell_\eta}$, [m^{-1}]
K	minimal spectral cut-off $\min\left(K_\nu, K_\eta\right)$, [m^{-1}]
$K_{\max}$	maximal spectral cut-off $\max\left(K_\nu, K_\eta\right)$, [m^{-1}]
ω	Fourier frequency, [s^{-1}]
ω_i	frequencies of Lehnert waves defined in (3.9), [s^{-1}]
$\Delta\omega$	frequency difference defined in (3.10), [s^{-1}]
$\mathbf{q}$	4-component spectral vector $\mathbf{q} = (\mathbf{k}, \omega)$
$\mathring{\Omega}$	solid angle in $\mathbf{k}$-space $\mathrm{d}\mathring{\Omega} = \sin\theta\mathrm{d}\theta\mathrm{d}\varphi$
X	cosine of polar angle in (k, θ, φ)-space $X = \cos\theta$
γ_ν	$\gamma_\nu = -\mathrm{i}\omega + \nu k^2$, [s^{-1}]
γ_η	$\gamma_\eta = -\mathrm{i}\omega + \eta k^2$, [s^{-1}]
γ_u	$\gamma_u = \gamma_\nu + \frac{(\mathbf{k}\cdot\langle\mathbf{B}\rangle)^2}{\gamma_\eta}$, [s^{-1}]
$\mathcal{E}$	electromotive force normalized by $\sqrt{\mu_0\rho}$, $\mathcal{E} = \langle\mathbf{u}'\times\mathbf{b}'\rangle$, cf (2.26), [m^2 s^{-2}]
$\boldsymbol{\alpha}$	the α-tensor, cf (2.26), [m s^{-1}]
$\boldsymbol{\beta}$	turbulent diffusion tensor, cf (2.26), [m^{-2} s^{-1}]
$\boldsymbol{\gamma}$	cross-helicity dynamo tensor (vortical), cf (2.26), [m^{-2} s^{-1}]
$\tilde{\boldsymbol{\gamma}}$	cross-helicity dynamo tensor (rotational), cf (2.31d), [m^{-2} s^{-1}]
$\mathbf{v}_{\mathrm{drift}}$	drift velocity (from $\boldsymbol{\alpha}^{(a)}$), cf (2.26), [m s^{-1}]

$\bar{\alpha}$	mean α-coefficient $\bar{\alpha} = \alpha = \mathrm{Tr}\,\boldsymbol{\alpha}/3$, [m s^{-1}]
$\breve{\alpha}$	fluctuational α-coefficient, cf (4.79b), [m s^{-1}]
α_X	α-coefficient from cross helicity, cf (5.51), [m s^{-1}]
α_R	α-coefficient from residual helicity, cf (5.50), [m s^{-1}]
α_{Xneq}	α-coefficient from joint action of cross and kinetic helicities, cf (5.67), [m s^{-1}]
$\langle \mathbf{A} \cdot \mathbf{B} \rangle$	magnetic helicity, cf section 2.1.1.1, [m^3 s^{-2}]
$\langle \mathbf{U} \cdot \mathbf{B} \rangle$	cross helicity, cf section 2.1.1.2, [m^2 s^{-2}]
$\langle \mathbf{U} \cdot \mathbf{W} \rangle$	kinetic helicity, cf section 2.1.1.3, [m s^{-2}]
$\langle \mathbf{B} \cdot \mathbf{J} \rangle$	current helicity, cf section 2.1.1.4, [m s^{-2}]
$\langle e_k \rangle$	turbulent kinetic energy $\langle e_k \rangle = \frac{1}{2}\langle u'^2 \rangle$, [m^2 s^{-2}]
$\langle e_m \rangle$	turbulent magnetic energy $\langle e_m \rangle = \frac{1}{2}\langle b'^2 \rangle$, [m^2 s^{-2}]
$\langle h_k \rangle$	turbulent kinetic helicity $\langle h_k \rangle = \langle \mathbf{u}' \cdot \mathbf{w}' \rangle$, [m s^{-2}]
$\langle h_r \rangle$	residual helicity $\langle h_r \rangle = -\langle \mathbf{u}' \cdot \mathbf{w}' \rangle + \langle \mathbf{b}' \cdot \mathbf{j}' \rangle$, cf section 2.1.1.4, [m s^{-2}]
Υ	nondimensional turbulent cross helicity, cf (5.63), [*nondimensional*]
$T_{ij}^{(s)}$	symmetric part of a tensor $\mathbf{T}$, $T_{ij}^{(s)} = \frac{1}{2}\left(T_{ij} + T_{ji}\right)$
$T_{ij}^{(a)}$	antisymmetric part of a tensor $\mathbf{T}$, $T_{ij}^{(a)} = \frac{1}{2}\left(T_{ij} - T_{ji}\right)$
$\mathbf{T}\colon \mathbf{S}$	double contraction $T_{ij}S_{ij}$
$P_{ij}(\mathbf{k})$	projection operator to the plane $\perp\mathbf{k}$, $P_{ij}(\mathbf{k}) = \delta_{ij} - \frac{k_i k_j}{k^2}$, [*nondimensional*]
$P_{imn}(\mathbf{k})$	$= k_m P_{in}(\mathbf{k}) + k_n P_{im}(\mathbf{k})$, [m^{-1}]
$M_{imn}(\mathbf{k})$	$= \frac{1}{2} P_{imn}(\mathbf{k})$, [m^{-1}]
$N_{ijk}(\mathbf{k})$	$= \frac{1}{2}\left(k_j \delta_{ik} - k_k \delta_{ij}\right)$, [m^{-1}]
$\mathfrak{G}_{ij}$	operator defined in (5.70), [s]
$G_{ij}^{(uu)}$, $G_{ij}^{(bb)}$, $G_{ij}^{(ub)}$, $G_{ij}^{(bu)}$	spectral Green tensors of MHD turbulence, cf section (5.1.1), [*nondimensional*]
$\mathfrak{Re}$	real part of a complex number
$\mathfrak{Im}$	imaginary part of a complex number
f^*	complex conjugate of f
c. c.	complex conjugate of a complex number
Pm	magnetic Prandtl number $Pm = \frac{\nu}{\eta}$, [*nondimensional*]
M	global Hartmann number $M = \frac{\langle B \rangle L}{\sqrt{\nu\eta}}$, [*nondimensional*]
Ro	'Rossby' number $Ro = \frac{\Vert \langle \mathbf{u}' \rangle \Vert^2}{\ell \Vert \mathbf{f} \Vert}$, [*nondimensional*]
E_ν	'Ekman' number $E_\nu = \frac{\nu \Vert \langle \mathbf{u}' \rangle \Vert}{\ell^2 \Vert \mathbf{f} \Vert}$, [*nondimensional*]
E_η	resistive 'Ekman' number $E_\eta = \frac{\eta \Vert \langle \mathbf{u}' \rangle \Vert}{\ell^2 \Vert \mathbf{f} \Vert}$, [*nondimensional*]

$\tilde{\beta}$	magnetic magnitude ratio $\tilde{\beta} = \frac{\lVert \langle \mathbf{B} \rangle \rVert}{\lVert \langle \mathbf{b}' \rangle \rVert}$, [*nondimensional*]
H	fluctuational magnitude ratio $H = \frac{\lVert \langle \mathbf{b}' \rangle \rVert^2}{\lVert \langle \mathbf{u}' \rangle \rVert^2}$, [*nondimensional*]
ε, δ	scale separation parameters $\delta = \varepsilon = \frac{\ell_e}{L}$, [*nondimensional*]
$\mathbf{\Omega}_0$	scaled rotation vector $\frac{\mathbf{\Omega}}{\delta}$, [s^{-1}]
$\mathbf{X}$	slow spatial variable $\mathbf{X} = \varepsilon \mathbf{x}$, [m]
T	slow time variable $T = \varepsilon t$, [s]
$\bar{\nabla}$	gradient with respect to the slow variable $\bar{\nabla} = \nabla_{\mathbf{X}}$, [m^{-1}]
θ_{K_i}	unit step function $\theta_{K_i} = \theta(K_i - k)$

Krzysztof A Mizerski

Chapter 1

The equations of magnetohydrodynamics

The derivation of the fundamental equations governing the flow of Newtonian, electrically conducting fluids and the coupled evolution of magnetic fields has been provided in many books, e.g. in Chandrasekhar (1961) [1], Roberts (1967) [2], Moffatt and Dormy (2019) [3] and many others. Therefore, here we only briefly recall the main points of the derivation for incompressible, constant-density (ρ = const) fluids, with the aim to keep the book self-consistent and set grounds for later chapters.

1.1 The Navier–Stokes equation—momentum balance

Let us denote the uniform density of the fluid by ρ and its velocity field by $\mathbf{U}(\mathbf{x}, t)$, where $\mathbf{x}$ is the position vector and t denotes time. The momentum per unit volume, in other words the momentum density is $\rho\mathbf{u}$. The variation of fluid's momentum in a volume V can be attributed only to two phenomena, that is either to momentum sources (or sinks, which can be thought off as negative sources) within the volume resulting from volume forces acting on the fluid (e.g. gravity, Lorentz force, Coriolis force, etc) or the flux of the momentum, which at least partly, is due to the flow of the medium. Therefore, the total variation of momentum in a fixed volume V can be expressed as follows

$$\begin{aligned}&\frac{\mathrm{d}}{\mathrm{d}t}\int_V \rho\mathbf{U}(\mathbf{x}, t)\mathrm{d}V\\&=\rho\int_V \frac{\partial\mathbf{U}(\mathbf{x}, t)}{\partial t}\mathrm{d}V=\int_V \mathbf{F}(\mathbf{x}, t)\mathrm{d}V-\oint_{\partial V}[\rho\mathbf{U}(\mathbf{x}, t)\mathbf{U}(\mathbf{x}, t)-\boldsymbol{\sigma}(\mathbf{x}, t)]\cdot\hat{\mathbf{n}}\mathrm{d}\Sigma,\end{aligned} \tag{1.1}$$

where $\hat{\mathbf{n}}$ is the unit normal directed outside of the surface ∂V enclosing the volume V, $\mathbf{F}(\mathbf{x}, t)$ is the body force density and $\rho\mathbf{UU}-\boldsymbol{\sigma}$ is the flux of momentum (tensorial, since the momentum is vectorial) with $\rho\mathbf{UU}$ being the advective flux and $\boldsymbol{\sigma}$ the stress tensor (including pressure and frictional flux). Therefore, momentum is conserved in a volume V if there are no body forces the same amount of flux enters and leaves the

doi:10.1088/978-0-7503-6362-4ch1

volume. Since V is arbitrary, by the use of the Gauss–Ostrogradsky divergence theorem one obtains,

$$\oint_{\partial V} [\rho\mathbf{U}(\mathbf{x}, t)\mathbf{U}(\mathbf{x}, t) - \boldsymbol{\sigma}(\mathbf{x}, t)] \cdot \hat{\mathbf{n}}\mathrm{d}\Sigma = \int_V \nabla \cdot [\rho\mathbf{U}(\mathbf{x}, t)\mathbf{U}(\mathbf{x}, t) - \boldsymbol{\sigma}(\mathbf{x}, t)]\mathrm{d}^3\mathbf{x}, \quad (1.2)$$

which by (1.1) leads to the local version of the Newton's second law of dynamics in the form

$$\frac{\partial \mathbf{U}}{\partial t} + \nabla \cdot \left(\mathbf{U}\mathbf{U} - \frac{\boldsymbol{\sigma}}{\rho}\right) = \mathbf{F}, \quad (1.3)$$

where we have dropped the notation with explicit dependence on $(\mathbf{x}, t)$ for clarity. The components of the stress tensor $\boldsymbol{\sigma}$ describe forces per unit surface that fluid elements exert on each other in a flow. More precisely a force per unit surface exerted on a given fluid element by a neighbouring element is given by $\boldsymbol{\sigma} \cdot \hat{\mathbf{n}}$, where $\hat{\mathbf{n}}$ is the unit normal to the surface separating the two fluid elements, directed towards the neighbour. The fundamental assumption for Newtonian fluids is that the dissipative part of the stress tensor, say $\boldsymbol{\sigma}_{\text{diss}}$, associated with frictional effects due to the nonuniform flow, is linearly related to the tensor of deformation rate, that is, the flow velocity gradient tensor

$$\mathbb{G}_{ij} = \frac{\partial U_i}{\partial x_j}, \quad (1.4)$$

since it is the presence of velocity gradients which is necessary and sufficient for frictional forces to appear; in the above and in the following the subscripts ij denote the cartesian components of a tensor. Once this assumption is made, the final form of the constitutive relation defining the stress tensor for an isotropic fluid is simply an outcome of symmetry of the stress tensor, which follows from the angular momentum balance (cf [4, 5] or [6]) and very basic properties of tensorial objects known from the group theory. The latter are simply contained in the Curie's principle 1894 (cf [7, 8]), which states that in an isotropic system only those tensorial objects which at rotations of a system of reference transform according to the same irreducible (therefore distinct) representations of the rotation groups can be linearly related. Since the trace and the symmetric traceless part of a tensor transform differently, one arrives at the constitutive relations

$$\mathrm{Tr}\,\boldsymbol{\sigma}_{\text{diss}} = 3\rho\nu_b\,\mathrm{Tr}\,\mathbb{G} = 3\rho\nu_b\nabla \cdot \mathbf{U}, \quad (1.5)$$

and

$$\boldsymbol{\sigma}_{\text{diss}} - \frac{1}{3}(\mathrm{Tr}\,\boldsymbol{\sigma}_{\text{diss}})\mathbf{I} = 2\rho\nu\left[\mathbb{G}^{(s)} - \frac{1}{3}(\nabla \cdot \mathbf{u})\mathbf{I}\right], \quad (1.6)$$

where the superfix (s) denotes the symmetric part of a tensor, such as

$$\mathbb{G}_{ij}^{(s)} = \frac{1}{2}\left(\frac{\partial U_i}{\partial x_j} + \frac{\partial U_j}{\partial x_i}\right), \quad (1.7)$$

Tr denotes the trace of a tensor, **I** is the unitary matrix and the coefficients of proportionality are the kinematic shear viscosity $\nu > 0$ and the so-called bulk viscosity $\nu_b > 0$ associated with compressional effects; these coefficients must be positive since they describe frictional, therefore dissipative effects. A great simplification is obtained, when the fluid can be assumed incompressible, when[1]

$$\mathrm{Tr}\,\mathbb{G} = \nabla \cdot \mathbf{U} = 0, \tag{1.10}$$

and in this book we will consider only fluids with uniform density. It follows that for incompressible, isotropic Newtonian fluids the stress tensor is expressed in terms of fluid pressure p and velocity gradients in the following way (cf e.g. [1, 4–6])

$$\frac{\sigma_{ij}}{\rho} = -\frac{p}{\rho}\delta_{ij} + 2\nu\mathbb{G}_{ij}^{(s)}. \tag{1.11}$$

The dissipative part of the stress tensor (1.11), that is

$$\sigma_{\mathrm{diss}\;ij} = 2\rho\nu\mathbb{G}_{ij}^{(s)} = \rho\nu\left(\frac{\partial U_i}{\partial x_j} + \frac{\partial U_j}{\partial x_i}\right), \tag{1.12}$$

describes the viscous shear stresses in the fluid. Consequently, under the assumption that there are no mass sources in the entire fluid volume, the fundamental momentum balance (1.3) takes the form

$$\frac{\partial \mathbf{U}}{\partial t} + (\mathbf{U} \cdot \nabla)\,\mathbf{U} = \frac{1}{\rho}\nabla \cdot \boldsymbol{\sigma} + \mathbf{F}, \tag{1.13}$$

and after introduction of the stress tensor form for a Newtonian, incompressible fluid (1.11), we arrive at the well-known Navier–Stokes equation,

$$\frac{\partial \mathbf{U}}{\partial t} + (\mathbf{U} \cdot \nabla)\,\mathbf{U} = -\nabla\frac{p}{\rho} + \mathbf{F} + \nu\nabla^2\mathbf{U}, \tag{1.14}$$

which describes the evolution of the velocity field in the flow of a fluid with uniform viscosity and density. This equation must be supplemented by the mass conservation equation, which in the incompressible case reduces to (1.10) and only then can both the velocity and pressure fields be determined.

---

[1] The law of mass conservation for a continuous medium can be obtained by consideration of a mass balance in a fixed volume V

$$\frac{\mathrm{d}}{\mathrm{d}t}\int_V \rho(\mathbf{x}, t)\mathrm{d}V = \int_V \frac{\partial\rho(\mathbf{x}, t)}{\partial t}\mathrm{d}V = \int_V S_\rho(\mathbf{x}, t)\mathrm{d}V - \oint_{\partial V}\rho(\mathbf{x}, t)\mathbf{u}(\mathbf{x}, t)\cdot\hat{\mathbf{n}}\mathrm{d}\Sigma, \tag{1.8}$$

where $S_\rho(\mathbf{x}, t)$ denotes mass sources in the volume V and since the flux of mass in a fluid results solely from the flow $\mathbf{u}(\mathbf{x}, t)$, it takes the form $\rho(\mathbf{x}, t)\mathbf{u}(\mathbf{x}, t)$. Hence locally the continuity equation reads

$$\frac{\partial\rho(\mathbf{x}, t)}{\partial t} + \nabla\cdot[\rho(\mathbf{x}, t)\mathbf{u}(\mathbf{x}, t)] = S_\rho(\mathbf{x}, t), \tag{1.9}$$

which expresses local mass conservation if the sources of matter $S_\rho(\mathbf{x}, t)$ vanish at least locally.

1.2 The induction equation—magnetic field evolution

Naturally, the forcing can also involve additional unknown fields, the evolution of which is coupled with the evolution of the fluid flow, as in the case of the magnetic Lorentz force exerted on the flow of an electrically conducting fluid in presence of magnetic fields. Denoting the magnetic field by $\mathbf{B}(\mathbf{x}, t)$ and the electric currents by $\mathbf{J}(\mathbf{x}, t)$ the Lorentz force per unit volume is

$$\mathbf{F}_{\mathrm{L}} = \mathbf{J} \times \mathbf{B} = \text{Lorentz force}, \tag{1.15}$$

i.e. it is the cross product of currents and the magnetic field. One may already notice, that such a form of the body force allows for force-free fields, when the currents flow along the magnetic lines of force, $\mathbf{J} \propto \mathbf{B}$, and the flow remains unaltered by the field, but the magnetic field, on the other hand, can be strongly influenced by the flow. Such situations are known to exist and have been studied since [9] and [10].

Therefore, the complete description of electrically conducting fluids requires one more equation determining the evolution of the magnetic field. The derivation involves the fundamental Maxwell laws of electromagnetism, supplied by the Ohm's law for electric currents, which is a material law of much less fundamental character, however, typically well satisfied in conducting liquids such as liquid metals and plasma. Because of the need for description of motion of a conducting fluid coupled with the magnetic field use will also be made of fundamental transformation laws for electromagnetic fields between the laboratory frame and one moving with the fluid's velocity $\mathbf{U}$. We start with Faraday's law of induction

$$\frac{\partial \mathbf{B}}{\partial t} = -\nabla \times \mathbf{E}, \tag{1.16}$$

which states that time-dependent magnetic fields are associated with circular (non-potential) electric fields, in other words, that a time-dependent flux of magnetic field through can induce an electromotive force and thus currents. The Ampère's law

$$\mu_0 \mathbf{J} = \nabla \times \mathbf{B} - \frac{1}{c^2}\frac{\partial \mathbf{E}}{\partial t}, \tag{1.17}$$

provides an expression for the current field, which is created by circular magnetic fields and time-dependent electric fields; the latter, $\partial_t \mathbf{E}/c^2$ is often called the displacement current. The constants $\mu_0 \approx 1.26 \times 10^{-6}\ \mathrm{N\,A^{-2}}$ and $c \approx 3 \times 10^8\ \mathrm{m\,s^{-1}}$ are the magnetic permittivity of vacuum and the speed of light, respectively. Gauss's law for magnetism

$$\nabla \cdot \mathbf{B} = 0, \tag{1.18}$$

simply states that there are no magnetic monopoles, and hence the flux of magnetic field through a closed surface is always zero. Only these three of the Maxwell equations will be needed here for description of magnetohydrodynamic flows, but in principle the last of the Maxwell equations, namely Gauss's law for the electric field can be used to calculate the charge distribution in the system, if necessary.

Next we turn to the Ohm's material law, which states that the current in a conducting material is proportional to the electric field which generates it and the coefficient of proportionality is called the electrical conductivity of the medium. However, for the case of a flow of conducting fluid this law can only be directly applied in the frame co-moving with the fluid's velocity **U**, and this is why we will need the Galilean transformation laws for electromagnetic fields to laboratory frame of reference. For an observer moving with the fluid

$$\vec{\mathbf{J}} = \sigma_e \vec{\mathbf{E}}, \tag{1.19}$$

where the upper arrows denote fields in the moving frame and σ_e is the electrical conductivity of the fluid, assumed constant. The non-relativistic laws of Galilean transformation between a laboratory frame and one moving with velocity **U** are (cf [11, 12])

$$\mathbf{B} = \vec{\mathbf{B}}, \quad \mathbf{J} = \vec{\mathbf{J}}, \quad \mathbf{E} = \vec{\mathbf{E}} - \mathbf{U} \times \mathbf{B}, \tag{1.20}$$

so that the magnetic field and the current field do not change under a Galilean transformation, but the electric field is altered by a contribution from the cross product $\mathbf{U} \times \mathbf{B}$; hence we can write

$$\mathbf{J} = \vec{\mathbf{J}} = \sigma_e \vec{\mathbf{E}}. \tag{1.21}$$

Substituting for the electric field **E** from (1.20) into Faraday's law (1.16) and then using Ohm's law (1.21) one obtains

$$\frac{\partial \mathbf{B}}{\partial t} = \nabla \times (\mathbf{U} \times \mathbf{B}) - \nabla \times \vec{\mathbf{E}} = \nabla \times (\mathbf{U} \times \mathbf{B}) - \frac{\nabla \times \mathbf{J}}{\sigma_e}. \tag{1.22}$$

Next substituting for **J** from the Ampère's law (1.17) leads to

$$\frac{\partial \mathbf{B}}{\partial t} = \nabla \times (\mathbf{U} \times \mathbf{B}) - \frac{1}{\mu_0 \sigma_e} \nabla \times \nabla \times \mathbf{B} + \frac{1}{\mu_0 \sigma_e} \frac{1}{c^2} \frac{\partial}{\partial t} \nabla \times \mathbf{E}, \tag{1.23}$$

and finally using again (1.16) we arrive at

$$\frac{\partial \mathbf{B}}{\partial t} = \nabla \times (\mathbf{U} \times \mathbf{B}) - \frac{1}{\mu_0 \sigma_e} \nabla \times \nabla \times \mathbf{B} - \frac{1}{\mu_0 \sigma_e c^2} \frac{\partial^2 \mathbf{B}}{\partial t^2}, \tag{1.24}$$

which involves the magnetic and velocity fields only and the electric field has been eliminated. The magnetic diffusivity is defined as

$$\eta = \frac{1}{\mu_0 \sigma_e}, \tag{1.25}$$

and will be assumed constant throughout. The vector identity $\nabla^2 \mathbf{B} = \nabla \nabla \cdot \mathbf{B} - \nabla \times \nabla \times \mathbf{B}$ and Gauss's law for magnetism (1.18) allow us to transform the latter equation into

$$\frac{\partial \mathbf{B}}{\partial t} = \nabla \times (\mathbf{U} \times \mathbf{B}) + \eta \nabla^2 \mathbf{B} - \frac{\eta}{c^2} \frac{\partial^2 \mathbf{B}}{\partial t^2}. \tag{1.26}$$

Next, we will show that the last term $\eta\partial_t^2\mathbf{B}/c^2$ resulting from displacement currents is typically negligible in laboratory and natural systems. Comparison of the terms with the first and second time derivative leads to

$$\frac{\|\partial_t\mathbf{B}\|}{\|\,\eta\partial_t^2\mathbf{B}/c^2\,\|} \sim \frac{c^2\mathcal{T}}{\eta} = \frac{c}{\mathcal{U}}\frac{c}{\eta/\mathcal{L}}, \tag{1.27}$$

where $\mathcal{T} = \mathcal{L}/\mathcal{U}$ is the inertial time scale of variation of the magnetic field and $\mathcal{U}$ is the typical velocity scale in the flow and $\mathcal{L}$ is the typical length scale of magnetic field variation. In all non-relativistic cases considered in this book, such as laboratory magnetohydrodynamic experiments, or flow dynamics in planetary cores and stellar interiors, the ratio

$$\frac{c}{\mathcal{U}} \gg 1, \tag{1.28}$$

is extremely large; for example, the flow velocities in the Earth's core have the magnitude of millimetres per second and the flow magnitudes in the Solar interior and laboratory experiments can be estimated at 1–10 m s^{-1}, and both estimates are indeed much smaller than the speed of light. The second ratio in (1.27), namely that of the speed of light c to the characteristic magnetic diffusion speed $\eta/\mathcal{L}$ is also typically greater than unity. For liquid iron (planetary cores) and liquid sodium (experiments) the magnetic diffusivity is roughly in the range 10^{-1} – 1 $m^2\,s^{-1}$, hence for very large length scales corresponding to size of the Earth's core, $\mathcal{L} \approx 10^6$ m, the characteristic diffusion velocity is very small $\eta/\mathcal{L} \approx 10^{-6}$ m s^{-1} and for laboratory length scales of a few meters the diffusion velocity is roughly at order unity. In the case of the Solar interior the $\eta \approx 10^{-2}$ $m^2\,s^{-1}$, hence for any typical length scales of magnetic field variation in the Sun the diffusion velocity is small. This shows that the displacement current is entirely negligible in the considered non-relativistic systems, and thus the term $\eta\partial_t^2\mathbf{B}/c^2$ in (1.26) will be neglected, as typically in the cases of systems for which the characteristic evolution times are much longer than the time required for electromagnetic waves to cross the system. This implies that we filter out the fast phenomena associated with propagation of electromagnetic waves, and thus the non-relativistic induction equation takes the form

$$\frac{\partial\mathbf{B}}{\partial t} = \nabla \times (\mathbf{U} \times \mathbf{B}) + \eta\nabla^2\mathbf{B}. \tag{1.29}$$

The second term on the right hand size describes diffusion of the magnetic energy, whereas the first one, by the use of the following vector identity satisfied for solenoidal fields ($\nabla \cdot \mathbf{U} = 0$, $\nabla \cdot \mathbf{B} = 0$)

$$\nabla \times (\mathbf{U} \times \mathbf{B}) = (\mathbf{B} \cdot \nabla)\mathbf{U} - (\mathbf{U} \cdot \nabla)\mathbf{B}, \tag{1.30}$$

describes advection, $(\mathbf{U} \cdot \nabla)\mathbf{B}$, and straining of the field lines, $(\mathbf{B} \cdot \nabla)\mathbf{U}$, by the flow. Under neglection of the diffusive effects the induction equation can be written in the form

$$\frac{\partial\mathbf{B}}{\partial t} = (\mathbf{B} \cdot \nabla)\mathbf{U} - (\mathbf{U} \cdot \nabla)\mathbf{B}, \tag{1.31}$$

with a commutator of the fields **U** and **B** on the right hand side. It follows, that the well-known Kelvin–Helmholtz, frozen-flux theorem applies, and similarly as in the case of vorticity lines in the flow of an inviscid fluid, the magnetic lines of force are frozen into the flow of a perfectly conducting liquid. Non-zero magnetic diffusivity leads to slipping of the lines relative to the velocity of the fluid. Thus the field lines are advected, stretched and bent by the flow, but the Lorentz force, according to the Lenz rule resists the deformations and its action on the flow eventually leads to some saturated state. This can be seen by rewriting the expression (1.15) for the Lorentz force in the following way

$$\mathbf{J} \times \mathbf{B} = \frac{1}{\mu_0}(\nabla \times \mathbf{B}) \times \mathbf{B} = -\nabla \frac{B^2}{2\mu_0} + \frac{1}{\mu_0}(\mathbf{B} \cdot \nabla)\mathbf{B}, \tag{1.32}$$

where we have used Ampère's law in the form

$$\mu_0 \mathbf{J} = \nabla \times \mathbf{B}, \tag{1.33}$$

i.e. with the displacement current neglected (in accordance with the above reasoning) and a vector identity. This shows that the Lorentz force consists of two contributions, the potential one, $-\nabla B^2/2\mu_0$ which is termed the magnetic pressure, and $(\mathbf{B} \cdot \nabla)\mathbf{B}$. The latter naturally tends to straighten back the magnetic field lines bent by the flow, that is any fold of the lines tends to be untwisted by the Lorentz force

$$\begin{aligned}(\mathbf{B} \cdot \nabla)\mathbf{B} = B\frac{\partial(B\hat{\mathbf{e}}_s)}{\partial s} &= \frac{\partial}{\partial s}\left(\frac{B^2}{2}\right)\hat{\mathbf{e}}_s + B^2\frac{\partial \hat{\mathbf{e}}_s}{\partial s} \\ &= \frac{\partial}{\partial s}\left(\frac{B^2}{2}\right)\hat{\mathbf{e}}_s + \frac{1}{\varrho}B^2\hat{\mathbf{e}}_n,\end{aligned} \tag{1.34}$$

where $\hat{\mathbf{e}}_s$ is the unit vector tangent to a field line, $\hat{\mathbf{e}}_n$ the unit normal (directed inwards, into the fold of the line), s is the parameter along the line (the arclength parameter), $\varrho > 0$ denotes the radius of curvature of the line and the Frenet–Serret formula $\mathrm{d}_s\hat{\mathbf{e}}_s = \hat{\mathbf{e}}_n/\varrho$ was used. The tendency to unfold the field lines can be seen on the normal component, which is positive definite and directed inwards, into the fold, so that the force acts to minimize the folding and straighten out the line, cf figure 1.1. The unstraining of the lines on the $\hat{\mathbf{e}}_s$ component takes place simultaneously.

The hydromagnetic dynamo action, i.e. spontaneous amplification of initially weak magnetic energy by the flow of an electrically conducting medium, in general terms, can be understood by inspection of the induction equation (1.29). As explained, the first term on the right hand side is responsible for advection, bending and stretching of the magnetic lines of force. These effects, through the Ampère's law (1.33), lead to formation of electric currents which in turn can lead to induction of a magnetic field. The dynamo effect can be suppressed by diffusion, if the latter is too strong. In order to quantify the influence of advection and straining relative to diffusion, we write down the induction equation in non-dimensional form

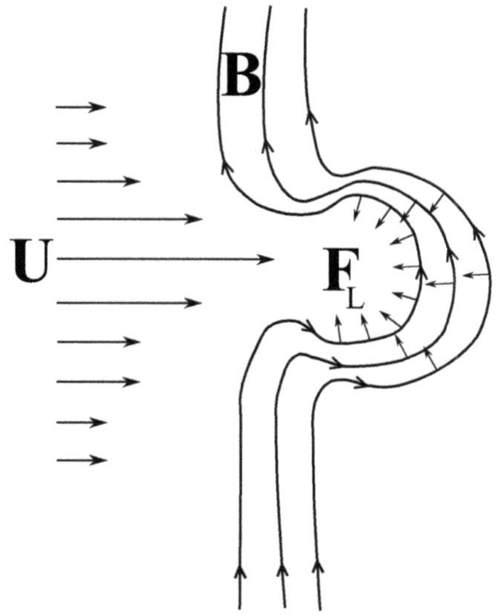

Figure 1.1. The action of the Lorentz force in nonuniform flow of a conducting fluid, which tends to straighten back the magnetic lines of force bent by the flow.

$$\frac{\partial \mathbf{B}}{\partial t} = \nabla \times (\mathbf{U} \times \mathbf{B}) + \frac{1}{Rm}\nabla^2\mathbf{B}, \tag{1.35}$$

where

$$Rm = \frac{\mathcal{UL}}{\eta}, \tag{1.36}$$

is a non-dimensional parameter termed the magnetic Reynolds number. When this parameter is large, the diffusive effects become weak compared to advection, bending and stretching of the field lines and dynamo action becomes possible. In the opposite case, $Rm \ll 1$, the dynamo action is unlikely, as diffusion dominates the dynamics. Not every flow **U** can induce dynamo action, and for example the famous Cowling's (1933) [13] theorem states that the hydromagnetic dynamo is not possible when both **U** and **B** have a common axis of symmetry. Similarly the Zel'dovich's (1957) [14] theorem excludes dynamo action by a two-dimensional flow.

An important distinction is that between kinematic and fully dynamic hydromagnetic dynamos. The former is a concept relying on entire neglection of the effect of the Lorentz force, i.e. kinematic dynamos are induced by a fixed flow, given beforehand, not interacting with the magnetic field. In the realistic, dynamic situation the back reaction of the Lorentz force on the flow plays an important role in the dynamo process, even in formation of the electromotive force. Although the problem of kinematic dynamos is certainly very important and allows for simple dynamo models, at the same time direct translation of intuitions developed on the basis of kinematic dynamos to the fully dynamic problem may be very misleading, since the former are linear in the magnetic field and hence dynamically very different from the full, nonlinear problem.

References

[1] Chandrasekhar S 1961 Hydrodynamic and hydromagnetic stability *International Series of Monographs on Physics* ed W Marshall and D H Wilkinson (Oxford: Clarendon Press (since 1981 printed by Dover Publications))

[2] Roberts P H 1967 *An Introduction to Magnetohydrodynamics* (London: Longmans Green)

[3] Moffatt H K and Dormy E 2019 *Self-exciting Fluid Dynamos* (Cambridge: Cambridge University Press)

[4] Landau L D and Lifshitz E M 1987 *Fluid mechanics Course of Theoretical Physics* **vol 6** (Oxford: Elsevier)

[5] Batchelor G K 1967 *An Introduction to Fluid Mechanics* (Cambridge: Cambridge University Press)

[6] Mizerski K A 2021 *Foundations of Convection with Density Stratification* (Cham: Springer)

[7] Chalmers A F 1970 Curie's principle *Br. J. Phil. Sci.* **21** 133–48

[8] de Groot S R and Mazur P 1984 *Non-equilibrium Thermodynamics* (New York: Dover)

[9] Chandrasekhar S 1956 On force-free magnetic fields *Proc. Natl Acad. Sci. USA* **42** 1–5

[10] Chandrasekhar S and Kendall P C 1957 On force-free magnetic fields *Astrophs. J.* **126** 457–60

[11] Jackson J D 1975 *Classical Electrodynamics* (New York: Wiley)

[12] Griffiths D J 1999 *Introduction to Electrodynamics* (Englewood Cliffs, NJ: Prentice-Hall)

[13] Cowling T G 1933 The magnetic field of sunspots *Mon. Not. R. Astron. Soc.* **94** 39–48

[14] Zel'dovich Ya B 1957 The magnetic field in the two-dimensional motion of a conducting turbulent fluid *Sov. Phys. JETP* **4** 460–2

IOP Publishing

Non-equilibrium Hydromagnetic Dynamos

Krzysztof A Mizerski

Chapter 2

The essentials of mean-field turbulent dynamo theory

We consider the evolution of a magnetic field in a viscous, electrically conducting fluid rotating with uniform angular velocity $\mathbf{\Omega}$. For simplicity, we suppose that the fluid is incompressible with constant uniform density ρ, magnetic diffusivity η and viscosity ν. Let $\mathbf{U}(\mathbf{x}, t)$ denote the velocity field and $\mathbf{B}(\mathbf{x}, t)$ the magnetic field in the fluid; then the standard magnetohydrodynamic equations governing the evolution of those fields take the following form

$$\frac{\partial \mathbf{U}}{\partial t} + (\mathbf{U} \cdot \nabla)\mathbf{U} = \mathbf{f} - \nabla\Pi - 2\mathbf{\Omega} \times \mathbf{U} + \frac{1}{\mu_0 \rho}(\mathbf{B} \cdot \nabla)\mathbf{B} + \nu\nabla^2\mathbf{U}, \tag{2.1a}$$

$$\frac{\partial \mathbf{B}}{\partial t} + (\mathbf{U} \cdot \nabla)\mathbf{B} = (\mathbf{B} \cdot \nabla)\mathbf{U} + \eta\nabla^2\mathbf{B}, \tag{2.1b}$$

$$\nabla \cdot \mathbf{U} = 0 \qquad \nabla \cdot \mathbf{B} = 0, \tag{2.1c}$$

where $\Pi = p/\rho + B^2/2\mu_0\rho - (\mathbf{\Omega} \times \mathbf{x})^2/2$ is the modified (total) pressure field, μ_0 is magnetic permeability of vacuum and $\mathbf{f}$ is a prescribed stirring force, independent of the magnetic, velocity and pressure fields. In the above the first equation is the well-known Navier–Stokes equation which, as explained in the previous chapter, represents the momentum balance for a fluid flow under action of the Coriolis force, $2\mathbf{\Omega} \times \mathbf{U}$, centrifugal force, $\mathbf{\Omega} \times (\mathbf{\Omega} \times \mathbf{x}) = \nabla(\mathbf{\Omega} \times \mathbf{x})^2/2$ (both being inertial forces resulting from background rotation, i.e. non-inertial frame of reference) and the magnetic Lorentz force, $\mathbf{J} \times \mathbf{B} = (\nabla \times \mathbf{B}) \times \mathbf{B}/\mu_0 = (\mathbf{B} \cdot \nabla)\mathbf{B}/\mu_0 - \nabla B^2/2\mu_0$, where $\mathbf{J}(\mathbf{x}, t)$ denotes the current density. According to Ampère's law

$$\mu_0 \mathbf{J} = \nabla \times \mathbf{B}, \tag{2.2}$$

doi:10.1088/978-0-7503-6362-4ch2

since in magnetohydrodynamics (MHD) the displacement currents (proportional to the temporal variations of the electric field) are neglected, because they correspond to very fast processes associated with electromagnetic waves traveling with the speed of light; such processes are negligible form the point of view of dynamo theory. The second equation (2.1b) is the induction equation for continuous medium in motion, resulting from Faraday's law of induction, Ampère's law (2.2) and Gauss's law for magnetism. The last two equations (2.1c) express the mass conservation law for an incompressible fluid, $\nabla \cdot \mathbf{U} = 0$, and the aforementioned Gauss's law, $\nabla \cdot \mathbf{B} = 0$. For a detailed derivation of the above system of equations the reader is directed to [1] and [2].

Without loss of generality we will consider solenoidal forcing

$$\nabla \cdot \mathbf{f} = 0, \tag{2.3}$$

since for the considered incompressible flows the potential part of a forcing only modifies the pressure. The background rotation is assumed uniform, so that

$$\mathbf{\Omega} = \Omega \hat{\mathbf{e}}_z, \qquad \Omega = \text{const}. \tag{2.4}$$

For the sake of simplicity we rescale the magnetic field $\mathbf{B}$ and the currents $\mathbf{J}$ in the following way

$$\frac{\mathbf{B}}{\sqrt{\mu_0 \rho}} \to \mathbf{B}, \qquad \sqrt{\frac{\mu_0}{\rho}}\mathbf{J} \to \mathbf{J}, \tag{2.5}$$

so that the former is expressed in velocity units and the latter in the units of vorticity and the prefactor $1/\mu_0\rho$ in the Lorentz-force term in the Navier–Stokes equation is lost; the equations now read

$$\frac{\partial \mathbf{U}}{\partial t} + (\mathbf{U} \cdot \nabla)\mathbf{U} = \mathbf{f} - \nabla \Pi - 2\mathbf{\Omega} \times \mathbf{U} + (\mathbf{B} \cdot \nabla)\mathbf{B} + \nu \nabla^2 \mathbf{U}, \tag{2.6a}$$

$$\frac{\partial \mathbf{B}}{\partial t} + (\mathbf{U} \cdot \nabla)\mathbf{B} = (\mathbf{B} \cdot \nabla)\mathbf{U} + \eta \nabla^2 \mathbf{B}, \tag{2.6b}$$

$$\nabla \cdot \mathbf{U} = 0 \qquad \nabla \cdot \mathbf{B} = 0, \tag{2.6c}$$

and Ampère's law is simply

$$\mathbf{J} = \nabla \times \mathbf{B}. \tag{2.7}$$

We shall consider the situation in which $\mathbf{U}$ and $\mathbf{B}$ are turbulent fields, spatially homogeneous on an ensemble of small length scales ℓ satisfying, $\ell_\nu, \ell_\eta \leqslant \ell \leqslant \ell_e$, where ℓ_e is the length scale of the most energetic (largest) turbulent eddies and ℓ_ν and ℓ_η are the viscous and resistive cut-offs, which in natural turbulence appear due to enhanced dissipation of energy at very small scales. In particular, in Kolmogorov-type turbulence $\ell_\nu = L^{1/4}(\nu/\| \langle U \rangle \|)^{3/4} = \ell_\eta Pm^{3/4}$, with L being the size of the entire system and $\| \langle U \rangle \|$ the large-scale velocity magnitude; the magnetic Prandtl number is defined as

$$Pm = \frac{\nu}{\eta}. \tag{2.8}$$

Next, let us denote by angular brackets the ensemble mean

$$\langle \cdot \rangle\text{–ensemble mean,}$$

and following the 'mean-field electrodynamics' approach of Steenbeck *et al* (1966) [3] we decompose **U** and **B** into mean and fluctuating parts

$$\mathbf{U} = \langle \mathbf{U} \rangle + \mathbf{u}', \quad \mathbf{B} = \langle \mathbf{B} \rangle + \mathbf{b}', \quad p = \langle p \rangle + p', \tag{2.9}$$

where $\langle \mathbf{u}' \rangle = \langle \mathbf{b}' \rangle = \mathbf{0}$, $\langle p' \rangle = 0$. We will treat $\langle \mathbf{B} \rangle$ as locally uniform, but varying weakly on scales $L \gg \ell_e$ much greater than that of the most energetic turbulent eddies. We will also neglect the large-scale component of the forcing $\langle \mathbf{f} \rangle$ and assume that the turbulence is stirred only at small scales

$$\langle \mathbf{f} \rangle = 0. \tag{2.10}$$

We can now write down separately the equations for the mean fields $\langle \mathbf{U} \rangle$ and $\langle \mathbf{B} \rangle$ and the turbulent fluctuations $\mathbf{u}'$ and $\mathbf{b}'$; this yields

$$\begin{aligned} \frac{\partial \langle \mathbf{U} \rangle}{\partial t} + (\langle \mathbf{U} \rangle \cdot \nabla)\langle \mathbf{U} \rangle = &- \nabla \langle \Pi \rangle - 2\mathbf{\Omega} \times \langle \mathbf{U} \rangle + (\langle \mathbf{B} \rangle \cdot \nabla)\langle \mathbf{B} \rangle + \nu \nabla^2 \langle \mathbf{U} \rangle \\ &- \nabla \cdot (\langle \mathbf{u}'\mathbf{u}' \rangle - \langle \mathbf{b}'\mathbf{b}' \rangle), \end{aligned} \tag{2.11a}$$

$$\frac{\partial \langle \mathbf{B} \rangle}{\partial t} = \nabla \times (\langle \mathbf{U} \rangle \times \langle \mathbf{B} \rangle) + \nabla \times \langle \mathbf{u}' \times \mathbf{b}' \rangle + \eta \nabla^2 \langle \mathbf{B} \rangle, \tag{2.11b}$$

$$\nabla \cdot \langle \mathbf{B} \rangle = 0, \quad \nabla \cdot \langle \mathbf{U} \rangle = 0. \tag{2.11c}$$

where

$$\boldsymbol{\mathcal{E}} = \langle \mathbf{u}' \times \mathbf{b}' \rangle, \tag{2.12}$$

is the large-scale electromotive force (EMF), whereas $\langle \mathbf{u}'\mathbf{u}' \rangle$ and $\langle \mathbf{b}'\mathbf{b}' \rangle$ are the so-called Reynolds and Maxwell stress tensors, respectively, which describe the net effect of the small-scale turbulent pulsations on the large-scale fields. The evolution of the former is governed by

$$\begin{aligned} &\frac{\partial \mathbf{u}'}{\partial t} - \nu \nabla^2 \mathbf{u}' + 2\mathbf{\Omega} \times \mathbf{u}' + (\langle \mathbf{U} \rangle \cdot \nabla)\mathbf{u}' - (\langle \mathbf{B} \rangle \cdot \nabla)\mathbf{b}' + \nabla \Pi' \\ &= \mathbf{f} - \nabla \cdot (\mathbf{u}'\mathbf{u}' - \mathbf{b}'\mathbf{b}' - \langle \mathbf{u}'\mathbf{u}' \rangle + \langle \mathbf{b}'\mathbf{b}' \rangle) - (\mathbf{u}' \cdot \nabla)\langle \mathbf{U} \rangle + (\mathbf{b}' \cdot \nabla)\langle \mathbf{B} \rangle, \end{aligned} \tag{2.13a}$$

$$\begin{aligned} \frac{\partial \mathbf{b}'}{\partial t} - \eta \nabla^2 \mathbf{b}' + (\langle \mathbf{U} \rangle \cdot \nabla)\mathbf{b}' - (\langle \mathbf{B} \rangle \cdot \nabla)\mathbf{u}' = &\nabla \times (\mathbf{u}' \times \mathbf{b}' - \langle \mathbf{u}' \times \mathbf{b}' \rangle) \\ &- (\mathbf{u}' \cdot \nabla)\langle \mathbf{B} \rangle + (\mathbf{b}' \cdot \nabla)\langle \mathbf{U} \rangle, \end{aligned} \tag{2.13b}$$

$$\nabla \cdot \mathbf{b}' = 0, \quad \nabla \cdot \mathbf{u}' = 0. \tag{2.13c}$$

Furthermore, in accordance with the above reasoning we assume scale separation between the means and the fluctuations, and express the dynamical fields using a Fourier transform defined in the following way

$$u_i'(\mathbf{x}, t) = \int_{K_e}^{K_\nu} \mathrm{d}^3k \int_{-\infty}^{\infty} \mathrm{d}\omega \hat{u}_i(\mathbf{k}, \omega)\mathrm{e}^{\mathrm{i}(\mathbf{k}\cdot\mathbf{x}-\omega t)}, \tag{2.14a}$$

$$b_i'(\mathbf{x}, t) = \int_{K_e}^{K_\eta} \mathrm{d}^3k \int_{-\infty}^{\infty} \mathrm{d}\omega \hat{b}_i(\mathbf{k}, \omega)\mathrm{e}^{\mathrm{i}(\mathbf{k}\cdot\mathbf{x}-\omega t)}, \tag{2.14b}$$

$$\langle U \rangle_i(\mathbf{x}, t) = \int_0^{\kappa_m} \mathrm{d}^3\kappa \int_{-\infty}^{\infty} \mathrm{d}\omega \, \langle \hat{U} \rangle_i(\boldsymbol{\kappa}, \omega)\mathrm{e}^{\mathrm{i}(\boldsymbol{\kappa}\cdot\mathbf{x}-\omega t)}, \tag{2.14c}$$

$$\langle B \rangle_i(\mathbf{x}, t) = \int_0^{\kappa_m} \mathrm{d}^3\kappa \int_{-\infty}^{\infty} \mathrm{d}\omega \, \langle \hat{B} \rangle_i(\boldsymbol{\kappa}, \omega)\mathrm{e}^{\mathrm{i}(\boldsymbol{\kappa}\cdot\mathbf{x}-\omega t)}, \tag{2.14d}$$

where the upper wavenumber cut-off for the mean-field Fourier modes κ_m and the lower wavenumber cut-off for the fluctuations $K_e = 2\pi/\ell_e$ associated with the most energetic eddies satisfy

$$\kappa_m \ll K_e. \tag{2.15}$$

The upper cut-offs for the Fourier modes in the velocity fluctuations $K_\nu = 2\pi/\ell_\nu$ and in the magnetic fluctuations $K_\eta = 2\pi/\ell_\eta$, as mentioned, appear due to viscous kinetic energy dissipation and resistive dissipation of magnetic energy. For generality the magnetic and kinetic cut-offs are assumed unequal, $K_\nu \neq K_\eta$. The assumption of separation of spatial scales is standard in mean-field theories, put forward in order to allow for analytic formulation of the dynamics of large-scale fields. Such a clear separation is often absent in natural systems and energy is present at all scales. Still, there are many cases when scale separation is very apparent. Hughes and Tobias (2010) [4] have estimated that a reasonable scale separation of approximately a decade between domain size and the size of the most energetic turbulent eddy, is achieved in numerical simulations with 1000^3 spectral modes, which accurately describe the dynamics at magnetic Reynolds numbers defined with the fluctuational velocity and length scales as $\sqrt{\langle \mathbf{u}'^2 \rangle}\,\ell/\eta$ of the order $\mathcal{O}(10^3)$. Even more importantly, natural scale separation is introduced by rapid background rotation in convective flows, such as, e.g. in the Earth's liquid core, where the small length scales are of the order $E^{1/3}L$, with $E = \nu/2\Omega L^2 \ll 1$ being the Ekman number and L the thickness of the convection zone. There are numerous examples of studies which utilize the multiple-scale approach for MHD turbulent convection, such as [5, 6], or more recently [7, 8] and [9]. In the mean-field theory a fluctuational wave vector $k = 2\pi/\ell$ is associated with the length scales ℓ regarded as the size of energetic eddies and $K_e = 2\pi/\ell_e$ is defined by the size of the largest, most energetic turbulent eddies, ℓ_e.

Note also, that since vector fields such as $\mathbf{u}'(\mathbf{x}, t)$ and $\mathbf{b}'(\mathbf{x}, t)$ are real, their Fourier transforms must satisfy

$$\hat{\mathbf{u}}(-\mathbf{k}, -\omega) = \hat{\mathbf{u}}^*(\mathbf{k}, \omega), \qquad \hat{\mathbf{b}}(-\mathbf{k}, -\omega) = \hat{\mathbf{b}}^*(\mathbf{k}, \omega), \tag{2.16}$$

where the star denotes the complex conjugate (and similarly for the $\langle\hat{\mathbf{U}}\rangle$ and $\langle\hat{\mathbf{B}}\rangle$ fields).

2.1 The turbulent electromotive force (EMF)

In the kinematic theory, i.e. when the turbulent flow is said to possess fixed statistical properties uninfluenced by the back reaction of the Lorentz force, a general formula for the mean EMF, $\boldsymbol{\mathcal{E}} = \langle\mathbf{u}' \times \mathbf{b}'\rangle$, can be obtained rather easily from the following symmetry considerations. First of all we observe from (cf (2.13b))

$$\begin{aligned}&\frac{\partial\mathbf{b}'}{\partial t} - \eta\nabla^2\mathbf{b}' + (\langle\mathbf{U}\rangle\cdot\nabla)\mathbf{b}' - (\mathbf{b}'\cdot\nabla)\langle\mathbf{U}\rangle - \nabla\times(\mathbf{u}'\times\mathbf{b}' - \langle\mathbf{u}'\times\mathbf{b}'\rangle)\\ &= -(\mathbf{u}'\cdot\nabla)\langle\mathbf{B}\rangle + (\langle\mathbf{B}\rangle\cdot\nabla)\mathbf{u}',\end{aligned} \tag{2.17}$$

that in the kinematic case the magnetic pulsation $\mathbf{b}'$ is a linear functional of $\langle\mathbf{B}\rangle$ and its gradient, $\nabla\langle\mathbf{B}\rangle$. However, in accordance with the assumed scale separation between the large-scale fields and the turbulent pulsations, turbulence is typically characterized by correlation length and time scales which are much shorter than the scales of variation of the mean fields such as $\langle\mathbf{B}\rangle$. Therefore, all correlation functions and in particular $\langle\mathbf{u}' \times \mathbf{b}'\rangle$ are non-zero only in a certain neighbourhood of a considered point in space-time $(\mathbf{x}, t)$, where the large-scale field can be sufficiently represented by $\langle\mathbf{B}\rangle(\mathbf{x})$ and its gradient. It follows that a sufficient representation of the mean EMF is obtained by a Taylor expansion of the linear functional $\mathcal{E}(\langle\mathbf{B}\rangle)$ up to first derivatives only and higher order derivatives are neglected (note that time derivatives of the mean field, $\partial_t\langle\mathbf{B}\rangle$ can always be replaced by its spatial derivatives from (2.11b))

$$\mathcal{E}_i = \alpha_{ij}\langle B\rangle_j + \kappa_{ijk}\frac{\partial\langle B\rangle_j}{\partial x_k}, \tag{2.18}$$

where α_{ij} and κ_{ijk} are pseudo tensors (skew tensors which change sign under reflections), because $\mathcal{E}$ is a polar vector and $\langle\mathbf{B}\rangle$ is skew; they are dependent on $\mathbf{u}'$ and $\langle\mathbf{U}\rangle$. The term $\alpha_{ij}\langle B\rangle_j$ in the mean EMF, which is often responsible for large-scale dynamo action in natural systems is commonly called the α-effect. The pseudo tensor κ_{ijk} can be split into symmetric and antisymmetric parts in the last two indices as

$$\kappa_{ijk}^{(s)} = \frac{1}{2}(\kappa_{ijk} + \kappa_{ikj}), \qquad \kappa_{ijk}^{(a)} = \frac{1}{2}(\kappa_{ijk} - \kappa_{ikj}), \tag{2.19}$$

which transforms (2.18) into

$$\boldsymbol{\mathcal{E}} = \boldsymbol{\alpha}\cdot\langle\mathbf{B}\rangle - \boldsymbol{\beta}\cdot\langle\mathbf{J}\rangle + \boldsymbol{\kappa}^{(s)}\colon(\nabla\langle\mathbf{B}\rangle)^{(s)}, \tag{2.20}$$

where the polar tensor

$$\beta_{ij} = \frac{1}{2}\kappa_{imn}\epsilon_{mnj}, \tag{2.21}$$

is typically called the turbulent diffusivity tensor and the double dot denotes contraction over two indices

$$(\boldsymbol{\kappa}^{(s)}: (\nabla\langle\mathbf{B}\rangle)^{(s)})_i = \kappa_{ijk}^{(s)}(\nabla\langle\mathbf{B}\rangle)_{jk}^{(s)}. \tag{2.22}$$

In a similar manner the α-effect is also sometimes split into two contributions, from the symmetric part $\alpha_{ij}^{(s)} = (\alpha_{ij} + \alpha_{ji})/2$ and the so-called effective drift velocity $\mathbf{v}_{\mathrm{drift}}$,

$$\mathrm{v}_{\mathrm{drift}\,i} = \frac{1}{2}\epsilon_{imn}\alpha_{nm}, \qquad \alpha_{ij}^{(a)} = \frac{1}{2}(\alpha_{ij} - \alpha_{ji}) = -\epsilon_{ijk}\mathrm{v}_{\mathrm{drift}\,k}, \tag{2.23}$$

which leads to

$$\boldsymbol{\mathcal{E}} = \boldsymbol{\alpha}^{(s)}\cdot\langle\mathbf{B}\rangle + \mathbf{v}_{\mathrm{drift}}\times\langle\mathbf{B}\rangle - \boldsymbol{\beta}\cdot\langle\mathbf{J}\rangle + \boldsymbol{\kappa}^{(s)}: (\nabla\langle\mathbf{B}\rangle)^{(s)}. \tag{2.24}$$

However, in the fully dynamical case, when the effect of the Lorentz force on the turbulent flow is significant, the mean EMF is in general a nonlinear functional of $\langle\mathbf{B}\rangle$ and $\langle\mathbf{U}\rangle$ and their gradients, $\nabla\langle\mathbf{B}\rangle$, $\nabla\langle\mathbf{U}\rangle$, including contributions from terms of the type $\langle\mathbf{B}\rangle\times\langle\mathbf{J}\rangle, \langle\mathbf{W}\rangle\times\langle\mathbf{J}\rangle, \nabla\cdot(\langle\mathbf{B}\rangle\langle\mathbf{B}\rangle)$, etc, where by $\mathbf{W} = \nabla\times\mathbf{U}$ we denote the vorticity. Inspection of equation (2.13b) leads to a conclusion that the dependence on the large-scale flow $\langle\mathbf{U}\rangle$, but not its gradient, can be easily removed due to the assumed scale separation between the large-scale fields and the turbulent pulsations by a simple Euclidean transformation to a local frame moving with velocity $\langle\mathbf{U}\rangle(\mathbf{x}, t)$ (in Fourier space this is done by a frequency shift, cf for example section 3.1). This is allowed, because of the fact that the correlations are short-range with respect to mean-field variational scales, hence in the neighbourhood of a space-time point $(\mathbf{x}, t)$ defined by the turbulent correlation length and time scales, the large-scale velocity can be sufficiently represented by $\langle\mathbf{U}\rangle(\mathbf{x})$ and its gradient only. If we further assume, that the mean magnetic field $\langle\mathbf{B}\rangle$ is weak, but still keeping the Lorentz force in, the mean EMF is obtained by a Taylor expansion of the functional $\boldsymbol{\mathcal{E}}(\langle\mathbf{B}\rangle, \langle\mathbf{U}\rangle)$ up to first derivatives, but the time derivatives of the mean flow are typically neglected. Therefore, the most general version of the expression for the turbulent EMF within the scope of the considered limits takes the form

$$\mathcal{E}_i = \alpha_{ij}\langle B\rangle_j + \kappa_{ijk}\frac{\partial\langle B\rangle_j}{\partial x_k} + \chi_{ijk}\frac{\partial\langle U\rangle_j}{\partial x_k}, \tag{2.25}$$

which in a similar manner as before can be transformed to

$$\begin{aligned}\boldsymbol{\mathcal{E}} = \langle\mathbf{u}'\times\mathbf{b}'\rangle = \boldsymbol{\alpha}^{(s)}\cdot\langle\mathbf{B}\rangle + \mathbf{v}_{\mathrm{drift}}\times\langle\mathbf{B}\rangle - \boldsymbol{\beta}\cdot\langle\mathbf{J}\rangle + \boldsymbol{\gamma}\cdot\langle\mathbf{W}\rangle \\ + \boldsymbol{\chi}^{(s)}: (\nabla\langle\mathbf{U}\rangle)^{(s)} + \boldsymbol{\kappa}^{(s)}: (\nabla\langle\mathbf{B}\rangle)^{(s)},\end{aligned} \tag{2.26}$$

where

$$\gamma_{ij} = \frac{1}{2}\chi_{imn}\epsilon_{nmj}. \tag{2.27}$$

It is worth mentioning that the expression (2.26) still holds true when the large-scale field $\langle\mathbf{B}\rangle$ is not weak, but turbulence is weak enough so that the nonlinear terms in the evolution equations for fluctuations (2.13a and b) of the type $\nabla \cdot (\mathbf{u}'\mathbf{u}')$ can be neglected, i.e. the evolution of the turbulent fluctuations can be considered linear; then the $\boldsymbol{\alpha}$ tensor can be a nonlinear functional of $\langle \mathbf{B} \rangle^2$. This is the case of the so-called weak turbulence, which we elaborate on in chapter 3. We should bear in mind, however, that as mentioned, in a general case of strong turbulence with strong magnetic fields the EMF, in principle, contains a number of additional contributions from terms nonlinear in the mean fields and their gradients and also from $\partial_t\langle \mathbf{U} \rangle$.

Let us shortly discuss some physical aspects associated with different contributions to the mean EMF in (2.26). Firstly we note that in the isotropic, statistically stationary and homogeneous case the α-effect was shown by [10] to be proportional to the difference between the current helicity $\langle \mathbf{b}' \cdot \mathbf{j}' \rangle$ and the kinetic helicity $\langle \mathbf{u}' \cdot \mathbf{w}' \rangle$, i.e. $\alpha \propto \langle -\mathbf{u}' \cdot \mathbf{w}' + \mathbf{b}' \cdot \mathbf{j}' \rangle$ (cf also e.g. [11] and [12]). Mizerski *et al* (2023) [13] have shown, that in non-equilibrium turbulence the cross-helicity $\langle \mathbf{u}' \cdot \mathbf{b}' \rangle$ and its dynamical history also plays an important role in generation of the α-effect, see chapter 5. The isotropic turbulent diffusivity is often estimated by the total turbulent energy, $\beta \propto \tau_t(\langle u'^2 \rangle + \langle b'^2 \rangle)/2$, where τ_t denotes the turn-over time of most energetic turbulent eddies, but in a most general case the β-effect includes also cross-helical contributions from $\langle \mathbf{u}' \cdot \mathbf{b}' \rangle^2$, cf [11, 12]. Furthermore, the $\boldsymbol{\gamma} \cdot \langle \mathbf{W} \rangle$ term represents the cross-helicity dynamo effect identified by [14] (see also [15] and a review by [16]), who utilized the *direct interaction approximation* introduced forty years earlier by Kraichnan (1959) [17] and argued that under the assumption of isotropy $\gamma \propto \langle \mathbf{u}' \cdot \mathbf{b}' \rangle$. When turbulence is not isotropic and inhomogeneous the antisymmetric part of the $\boldsymbol{\alpha}$-tensor, i.e. the effective drift velocity, $\mathbf{v}_{\mathrm{drift}}$, contributes to the variety of large-scale dynamo mechanisms; in fact the influence of compressibility effects on the EMF has been shown by [18] (cf equation (4.13) in this work) to significantly affect the drift velocity. There are even more effects associated with inhomogeneity of the large-scale fields which have been identified and result from the antisymmetric part of the $\boldsymbol{\beta}$-tensor. Those are the $\boldsymbol{\Omega} \times \langle \mathbf{J} \rangle$-effect[1] and the so-called shear-current effect, $\langle \mathbf{W} \rangle \times \langle \mathbf{J} \rangle$, see [19–21] and [22], the latter effect, however, is a second order effect in terms of the large-scale gradients, $\langle \mathbf{W} \rangle \times \langle \mathbf{J} \rangle = \mathcal{O}(\kappa_m^2 \ell_e^2)$, hence is typically neglected. The non-equilibrium effects which are the main subject of this book, modify all the existing effects and are largely independent of the magnetic diffusivity η.

As said, in strong MHD turbulence with strong mean magnetic field the situation is complex and the EMF depends on $\langle \mathbf{B} \rangle$ in a nonlinear way, e.g. contributions from $\langle \mathbf{B} \rangle \times \langle \mathbf{J} \rangle$, $\langle \mathbf{W} \rangle \times \langle \mathbf{J} \rangle$, $\nabla \cdot (\langle \mathbf{B} \rangle \langle \mathbf{B} \rangle)$ etc may appear. Nevertheless, as pointed out in [16] some general considerations for the fully dynamic, strong MHD turbulence can still be made. First, we rearrange (2.13a and b) into

$$\begin{aligned}\frac{\mathrm{D}\mathbf{u}'}{\mathrm{D}t} = \mathbf{f} - \nabla \Pi' - 2\boldsymbol{\Omega} \times \mathbf{u}' - (\mathbf{u}' \cdot \nabla)\langle \mathbf{U} \rangle + (\langle \mathbf{B} \rangle \cdot \nabla)\mathbf{b}' + (\mathbf{b}' \cdot \nabla)\langle \mathbf{B} \rangle \\ + \nu \nabla^2 \mathbf{u}' + \nabla \cdot (\mathbf{b}'\mathbf{b}') + \nabla \cdot (\langle \mathbf{u}'\mathbf{u}' \rangle - \langle \mathbf{b}'\mathbf{b}' \rangle),\end{aligned} \tag{2.28a}$$

[1] In general, differential rotation is possible $\boldsymbol{\Omega} = \boldsymbol{\Omega}(\mathbf{x})$, potentially making action of the term $\nabla \times (\boldsymbol{\Omega}(\mathbf{x}) \times \langle \mathbf{J} \rangle)$ even more efficient.

$$\frac{\mathrm{D}\mathbf{b}'}{\mathrm{D}t} = (\langle\mathbf{B}\rangle\cdot\nabla)\mathbf{u}' - (\mathbf{u}'\cdot\nabla)\langle\mathbf{B}\rangle + (\mathbf{b}'\cdot\nabla)\langle\mathbf{U}\rangle + (\mathbf{b}'\cdot\nabla)\mathbf{u}' + \eta\nabla^2\mathbf{b}' - \nabla\times\boldsymbol{\mathcal{E}}, \quad (2.28\text{b})$$

where

$$\frac{\mathrm{D}}{\mathrm{D}t} = \frac{\partial}{\partial t} + \mathbf{U}\cdot\nabla = \frac{\partial}{\partial t} + \langle\mathbf{U}\rangle\cdot\nabla + \mathbf{u}'\cdot\nabla, \quad (2.29)$$

is the Lagrangian derivative, and then derive the evolution equation for the mean EMF

$$\begin{aligned}\frac{\mathrm{D}\mathcal{E}_i}{\mathrm{D}t} =& \left\langle -u'_m\epsilon_{mik}\frac{\partial u'_k}{\partial x_j} + b'_m\epsilon_{mik}\frac{\partial b'_k}{\partial x_j}\right\rangle\langle B\rangle_j - \left(\left\langle u'_j u'_m\right\rangle + \left\langle b'_j b'_m\right\rangle\right)\epsilon_{imk}\frac{\partial\langle B\rangle_k}{\partial x_j} \\ &+ \left(\left\langle u'_j b'_m\right\rangle + \left\langle u'_m b'_j\right\rangle\right)\epsilon_{imk}\frac{\partial\langle U\rangle_k}{\partial x_j} + 2\Omega_i\left\langle u'_j b'_j\right\rangle - 2\Omega_j\left\langle u'_i b'_j\right\rangle \\ &+ \nu\partial_i\langle\mathbf{w}'\cdot\mathbf{b}'\rangle - \eta\partial_i\langle\mathbf{u}'\cdot\mathbf{j}'\rangle - \nu\partial_j\left\langle w'_i b'_j\right\rangle + \eta\partial_j\left\langle j'_i u'_j\right\rangle \\ &- \epsilon_{ijk}\frac{\partial\langle\Pi' b'_k\rangle}{\partial x_j} + \left\langle\Pi' j'_i\right\rangle + \epsilon_{ijk}\left\langle u'_j\frac{\partial(u'_k b'_m)}{\partial x_m}\right\rangle - \epsilon_{ijk}\left\langle b'_j\frac{\partial(b'_k b'_m)}{\partial x_m}\right\rangle \\ &+ \epsilon_{ijk}\left\langle f_j b'_k\right\rangle - \epsilon_{ijk}\frac{\partial}{\partial x_m}\left[\left\langle u'_m u'_j b'_k\right\rangle - u'_m\left\langle u'_j b'_k\right\rangle\right].\end{aligned} \quad (2.30)$$

On defining

$$\mathring{\alpha}_{ij} = \left\langle -u'_m\epsilon_{mik}\frac{\partial u'_k}{\partial x_j} + b'_m\epsilon_{mik}\frac{\partial b'_k}{\partial x_j}\right\rangle, \quad (2.31\text{a})$$

$$\mathring{\kappa}_{ijk} = \left\langle u'_j u'_m\right\rangle\epsilon_{imk} + \left\langle b'_j b'_m\right\rangle\epsilon_{imk}, \quad (2.31\text{b})$$

$$\mathring{\chi}_{ijk} = \left\langle u'_j b'_m\right\rangle\epsilon_{imk} + \left\langle u'_m b'_j\right\rangle\epsilon_{imk}, \quad (2.31\text{c})$$

$$\mathring{\gamma}_{ij} = \langle\mathbf{u}'\cdot\mathbf{b}'\rangle\delta_{ij} - \left\langle u'_i b'_j\right\rangle, \quad (2.31\text{d})$$

we can split those tensors into symmetric and antisymmetric parts in the last two indices in the following way

$$\mathring{\alpha}_{ij}^{(s)} = \frac{1}{2}\left(\mathring{\alpha}_{ij} + \mathring{\alpha}_{ji}\right), \quad \mathring{\alpha}_{ij}^{(a)} = \frac{1}{2}\left(\mathring{\alpha}_{ij} - \mathring{\alpha}_{ji}\right), \quad (2.32\text{a})$$

$$\mathring{\kappa}_{ijk}^{(s)} = \frac{1}{2}\left(\mathring{\kappa}_{ijk} + \mathring{\kappa}_{ikj}\right), \quad \mathring{\kappa}_{ijk}^{(a)} = \frac{1}{2}\left(\mathring{\kappa}_{ijk} - \mathring{\kappa}_{ikj}\right), \quad (2.32\text{b})$$

$$\mathring{\chi}_{ijk}^{(s)} = \frac{1}{2}\left(\mathring{\chi}_{ijk} + \mathring{\chi}_{ikj}\right), \quad \mathring{\chi}_{ijk}^{(a)} = \frac{1}{2}\left(\mathring{\chi}_{ijk} - \mathring{\chi}_{ikj}\right), \quad (2.32\text{c})$$

$$\mathring{\gamma}_{ij}^{(s)} = \frac{1}{2}\left(\mathring{\gamma}_{ij} + \mathring{\gamma}_{ji}\right), \quad \mathring{\gamma}_{ij}^{(a)} = \frac{1}{2}\left(\mathring{\gamma}_{ij} - \mathring{\gamma}_{ji}\right). \tag{2.32d}$$

Next, using $A_{ij} - A_{ji} = \epsilon_{ijk}\epsilon_{kmn}A_{mn}$ we arrive at

$$\mathring{\alpha}_{ij}^{(a)} = \frac{1}{2}\epsilon_{ijk}\frac{\partial}{\partial x_m}(\langle u_k' u_m'\rangle - \langle b_k' b_m'\rangle) \tag{2.33a}$$

$$\mathring{\kappa}_{ijk}^{(a)} = \frac{1}{2}\left(\langle u'^2\rangle + \langle b'^2\rangle\right)\epsilon_{ijk} - \frac{1}{2}(\langle u_i' u_m'\rangle + \langle b_i' b_m'\rangle)\epsilon_{mjk} \tag{2.33b}$$

$$\mathring{\chi}_{ijk}^{(a)} = \epsilon_{ijk}\langle \mathbf{u}' \cdot \mathbf{b}'\rangle - \frac{1}{2}(\langle u_i' b_m'\rangle + \langle b_i' u_m'\rangle)\epsilon_{mjk} \tag{2.33c}$$

$$\mathring{\gamma}_{ij}^{(a)} = -\frac{1}{2}\epsilon_{ijk}\mathcal{E}_k. \tag{2.33d}$$

The first few terms in (2.30) are

$$\begin{aligned}\frac{\mathrm{D}\boldsymbol{\mathcal{E}}}{\mathrm{D}t} &= \mathring{\boldsymbol{\alpha}}^{(s)} \cdot \langle \mathbf{B}\rangle + \mathring{\mathbf{v}}_{\mathrm{drift}} \times \langle \mathbf{B}\rangle - \frac{1}{2}\left(\langle u'^2\rangle + \langle b'^2\rangle\right)\langle \mathbf{J}\rangle + \frac{1}{2}(\langle \mathbf{u}'\mathbf{u}'\rangle + \langle \mathbf{b}'\mathbf{b}'\rangle) \cdot \langle \mathbf{J}\rangle \\ &+ \langle \mathbf{u}' \cdot \mathbf{b}'\rangle\langle \mathbf{W}\rangle - \frac{1}{2}(\langle \mathbf{u}'\mathbf{b}'\rangle + \langle \mathbf{b}'\mathbf{u}'\rangle) \cdot \langle \mathbf{W}\rangle + 2\mathring{\boldsymbol{\gamma}}^{(s)} \cdot \boldsymbol{\Omega} - \boldsymbol{\Omega} \times \boldsymbol{\mathcal{E}} \\ &+ \mathring{\boldsymbol{\chi}}^{(s)} : (\nabla\langle \mathbf{U}\rangle)^{(s)} - \mathring{\boldsymbol{\kappa}}^{(s)} : (\nabla\langle \mathbf{B}\rangle)^{(s)} + \ldots\end{aligned} \tag{2.34}$$

where we have defined

$$\mathring{\mathbf{v}}_{\mathrm{drift}} = -\frac{1}{2}\nabla \cdot (\langle \mathbf{u}'\mathbf{u}'\rangle - \langle \mathbf{b}'\mathbf{b}'\rangle), \tag{2.35}$$

and the dots indicate additional contributions from diffusion and the nonlinearities which, in principle, may be dynamically important, but are typically neglected due to high complexity. In the first few terms visible in (2.34) we recognize contributions of the same type as those obtained earlier in (2.26), in particular the α-effect is clearly still present in the fully nonlinear case.

Furthermore, we may put forward a simplifying assumption of homogeneity and isotropy of the turbulent flow, which implies $\boldsymbol{\Omega} = 0$ and also that any correlation function for turbulent fluctuations is expressed in the following manner

$$\left\langle f_i(\mathbf{x}, t) g_j(\mathbf{x}, t')\right\rangle = a_{\hat{\mathbf{r}}}(t, t')\delta_{ij} + b_{\hat{\mathbf{r}}}(t, t')\hat{r}_i\hat{r}_j + c_{\hat{\mathbf{r}}}(t, t')\epsilon_{ijk}\hat{r}_k, \tag{2.36}$$

where f and g denote either u or b, $\hat{\mathbf{r}} = \lim_{\mathbf{x}'\to\mathbf{x}}(\mathbf{x}' - \mathbf{x})/|\mathbf{x}' - \mathbf{x}|$ is *any* unit vector and the functions of t and t', $a_{\hat{\mathbf{r}}}(t, t')$, $b_{\hat{\mathbf{r}}}(t, t')$ and $c_{\hat{\mathbf{r}}}(t, t')$ are given by

$$\begin{aligned}b_{\hat{\mathbf{r}}}(t, t')\left(\hat{r}_i\hat{r}_j - \frac{1}{3}\delta_{ij}\right) &= \frac{1}{2}\left(\left\langle f_i(\mathbf{x}, t) g_j(\mathbf{x}, t')\right\rangle + \left\langle f_j(\mathbf{x}, t) g_i(\mathbf{x}, t')\right\rangle\right) \\ &- \frac{1}{3}\langle \mathbf{f}(\mathbf{x}, t) \cdot \mathbf{g}(\mathbf{x}, t')\rangle\delta_{ij},\end{aligned} \tag{2.37a}$$

$$a_{\hat{\mathbf{r}}}(t, t') = \frac{1}{3}\langle \mathbf{f}(\mathbf{x}, t) \cdot \mathbf{g}(\mathbf{x}, t') \rangle - \frac{1}{3} b_{\hat{\mathbf{r}}}(t, t'), \tag{2.37b}$$

$$c_{\hat{\mathbf{r}}}(t, t') = \frac{1}{2} \hat{\mathbf{r}} \cdot \langle \mathbf{f} \times \mathbf{g} \rangle. \tag{2.37c}$$

In such a case we get

$$\mathring{\alpha}_{ij}^{(s)} = \frac{1}{3}(-\langle \mathbf{u}' \cdot \mathbf{w}' \rangle + \langle \mathbf{b}' \cdot \mathbf{j}' \rangle)\delta_{ij}, \tag{2.38a}$$

$$\mathring{\mathbf{v}}_{\mathrm{drift}} = -\frac{1}{3}\nabla\left(\frac{\langle u'^2 \rangle}{2} - \frac{\langle b'^2 \rangle}{2}\right) = 0, \tag{2.38b}$$

$$\mathring{\boldsymbol{\kappa}}^{(s)} = 0, \qquad \mathring{\chi}^{(s)} = 0, \tag{2.38c}$$

$$\mathring{\gamma}_{ij}^{(s)} = \frac{2}{3}\langle \mathbf{u}' \cdot \mathbf{b}' \rangle \delta_{ij}, \tag{2.38d}$$

and the evolution equation for the turbulent EMF takes on a simpler form

$$\begin{aligned} \frac{\mathrm{D}\boldsymbol{\mathcal{E}}}{\mathrm{D}t} = &\frac{1}{3}(-\langle \mathbf{u}' \cdot \mathbf{w}' \rangle + \langle \mathbf{b}' \cdot \mathbf{j}' \rangle)\langle \mathbf{B} \rangle - \frac{1}{3}\left(\langle u'^2 \rangle + \langle b'^2 \rangle\right)\langle \mathbf{J} \rangle \\ &+ \frac{2}{3}\langle \mathbf{u}' \cdot \mathbf{b}' \rangle \langle \mathbf{W} \rangle + \ldots \end{aligned} \tag{2.39}$$

In stationary turbulence the solution is often crudely approximated by

$$\begin{aligned} \boldsymbol{\mathcal{E}} \approx &\frac{1}{3}\tau_t(-\langle \mathbf{u}' \cdot \mathbf{w}' \rangle + \langle \mathbf{b}' \cdot \mathbf{j}' \rangle)\langle \mathbf{B} \rangle - \frac{1}{3}\tau_t\left(\langle u'^2 \rangle + \langle b'^2 \rangle\right)\langle \mathbf{J} \rangle \\ &+ \frac{2}{3}\tau_t\langle \mathbf{u}' \cdot \mathbf{b}' \rangle \langle \mathbf{W} \rangle + \ldots \end{aligned} \tag{2.40}$$

and likewise one can crudely estimate $\mathbf{v}_{\mathrm{drift}} = \tau_t \mathring{\mathbf{v}}_{\mathrm{drift}}$, where τ_t denotes the turn-over time of most energetic turbulent eddies. Such estimates are not valid when turbulence is out of equilibrium, and as we will show in the following chapters additional contributions to the mean EMF arise from non-vanishing mean interactions between waves (Fourier modes) with distinct but close frequencies (beating waves) and from evolutional history of the product of kinetic and cross helicities. Although non-equilibrium effects influence all the contributions to the mean EMF described above, here we will concentrate solely on the influence of non-stationarity on the α-effect.

We can divide the dynamo effects, that is, the various contributions to the EMF, into those directly dependent on large-scale gradients and those indirectly dependent or independent of them. The α-effect (together with $\mathbf{v}_{\mathrm{drift}}$) is in the latter category, although naturally the α-effect is related to the helicities (such as the residual helicity $-\langle \mathbf{u}' \cdot \mathbf{w}' \rangle + \langle \mathbf{b}' \cdot \mathbf{j}' \rangle$ and the cross-helicity $\langle \mathbf{u}' \cdot \mathbf{b}' \rangle$) and as we demonstrate below the production of the helicities is in great part controlled by the large-scale shear and currents (indirect dependence). The α-effect can be further divided into diffusively controlled contributions (non-zero only when $\eta \neq 0$) and those from non-equilibrium effects, which are non-zero even in ideal conductors when $\eta = 0$.

Since the helicities of the turbulent flow are clearly crucial for the large-scale dynamo process to occur, we now turn to the problem of their evolution and production in the MHD turbulence.

2.1.1 Production of fluctuational helicities

In order to obtain and discuss the evolution equations for the kinetic, magnetic, current and cross helicities we must supply the dynamical equations governing the evolution of the magnetic vector potential defined by

$$\mathbf{B} = \nabla \times \mathbf{A}, \tag{2.41}$$

the vorticity $\mathbf{W} = \nabla \times \mathbf{U}$ and the electric current $\mathbf{J} = \nabla \times \mathbf{B}$, derived from (2.6a and b)

$$\frac{\partial \mathbf{A}}{\partial t} = \mathbf{U} \times \nabla \times \mathbf{A} - \nabla\phi + \eta\nabla^2\mathbf{A}, \tag{2.42a}$$

$$\frac{\partial \mathbf{W}}{\partial t} + (\mathbf{U} \cdot \nabla)\mathbf{W} = \nabla \times \mathbf{f} + 2(\mathbf{\Omega} \cdot \nabla)\mathbf{U} + (\mathbf{W} \cdot \nabla)\mathbf{U} + \nabla \times (\mathbf{J} \times \mathbf{B}) + \nu\nabla^2\mathbf{W}, \tag{2.42b}$$

$$\frac{\partial \mathbf{J}}{\partial t} = \nabla \times \nabla \times (\mathbf{U} \times \mathbf{B}) + \eta\nabla^2\mathbf{J}, \tag{2.42c}$$

where $\phi(\mathbf{x}, t)$ is some scalar function. Just as (2.28a and b), the equations which govern the evolution of fluctuations of $\mathbf{A}$, $\mathbf{W}$ and $\mathbf{J}$ can be obtained and cast in the following form

$$\begin{aligned}\frac{\mathrm{D}\mathbf{a}'}{\mathrm{D}t} = {} & \langle U\rangle_j \frac{\partial a_j'}{\partial x_i}\hat{\mathbf{e}}_i + u_j'\frac{\partial(\langle A\rangle_j + a_j')}{\partial x_i}\hat{\mathbf{e}}_i - \left\langle u_j'\frac{\partial a_j'}{\partial x_i}\right\rangle\hat{\mathbf{e}}_i \\ & - \nabla\phi' - (\mathbf{u}' \cdot \nabla)\langle\mathbf{A}\rangle + \langle(\mathbf{u}' \cdot \nabla)\mathbf{a}'\rangle + \eta\nabla^2\mathbf{a}',\end{aligned} \tag{2.43}$$

$$\begin{aligned}\frac{\mathrm{D}\mathbf{w}'}{\mathrm{D}t} = {} & \nabla \times \mathbf{f} + 2(\mathbf{\Omega} \cdot \nabla)\mathbf{u}' + (\langle\mathbf{W}\rangle \cdot \nabla)\mathbf{u}' - (\mathbf{u}' \cdot \nabla)\langle\mathbf{W}\rangle + (\mathbf{w}' \cdot \nabla)\langle\mathbf{U}\rangle \\ & + \nabla \times (\langle\mathbf{J}\rangle \times \mathbf{b}') + \nabla \times (\mathbf{j}' \times \langle\mathbf{B}\rangle) + (\mathbf{w}' \cdot \nabla)\mathbf{u}' - \langle(\mathbf{w}' \cdot \nabla)\mathbf{u}'\rangle \\ & + \nabla \times (\mathbf{j}' \times \mathbf{b}') - \nabla \times \langle\mathbf{j}' \times \mathbf{b}'\rangle + \langle(\mathbf{u}' \cdot \nabla)\mathbf{w}'\rangle + \nu\nabla^2\mathbf{w}',\end{aligned} \tag{2.44}$$

$$\begin{aligned}\frac{\mathrm{D}\mathbf{j}'}{\mathrm{D}t} = {} & \nabla \times \nabla \times (\langle\mathbf{U}\rangle \times \mathbf{b}' + \mathbf{u}' \times \langle\mathbf{B}\rangle + \mathbf{u}' \times \mathbf{b}') + [(\langle\mathbf{U}\rangle + \mathbf{u}') \cdot \nabla]\mathbf{j}' \\ & - \nabla \times \nabla \times \langle\mathbf{u}' \times \mathbf{b}'\rangle + \eta\nabla^2\mathbf{j}'.\end{aligned} \tag{2.45}$$

2.1.1.1 Magnetic helicity

The mean magnetic helicity has two contributions from the mean and fluctuational fields,

$$\langle \mathbf{A} \cdot \mathbf{B} \rangle = \langle \mathbf{A} \rangle \cdot \langle \mathbf{B} \rangle + \langle \mathbf{a}' \cdot \mathbf{b}' \rangle. \tag{2.46}$$

The total magnetic helicity $\mathcal{H}_m = \int_V \mathbf{A} \cdot \mathbf{B} \mathrm{d}V$ in the entire fluid volume V satisfies the following evolution equation, derived from (2.6b) and (2.42a)

$$\begin{aligned}\frac{\mathrm{D}}{\mathrm{D}t}\int_V \mathbf{A} \cdot \mathbf{B}\mathrm{d}V &= \int_V \{[\mathbf{U} \cdot (\nabla \mathbf{A})] \cdot \mathbf{B} + \mathbf{A} \cdot (\mathbf{B} \cdot \nabla)\mathbf{U} \\ &\quad - \nabla\phi \cdot \mathbf{B} + \eta\nabla^2\mathbf{A} \cdot \mathbf{B} + \eta\mathbf{A} \cdot \nabla^2\mathbf{B}\}\mathrm{d}V \\ &= \int_V \nabla \cdot [(\mathbf{A} \cdot \mathbf{U} - \phi + \eta\nabla \cdot \mathbf{A})\mathbf{B} + \eta\mathbf{A} \times \mathbf{J}]\mathrm{d}V \\ &\quad - 2\eta\int_V \mathbf{B} \cdot \mathbf{J}\mathrm{d}V.\end{aligned} \tag{2.47}$$

Therefore, it is always conserved in the absence of fluid's resistivity and magnetic flux through boundaries of the domain (cf [2]),

$$\frac{\mathrm{D}}{\mathrm{D}t}\int_V \mathbf{A} \cdot \mathbf{B}\mathrm{d}V = 0 \Rightarrow \int_V \langle \mathbf{A} \rangle \cdot \langle \mathbf{B} \rangle \mathrm{d}V = \mathcal{H}_m - \int_V \langle \mathbf{a}' \cdot \mathbf{b}' \rangle \mathrm{d}V, \quad \text{when } \eta = 0. \tag{2.48}$$

Naturally neither, the total magnetic helicity of the mean field $\int_V \langle \mathbf{A} \rangle \cdot \langle \mathbf{B} \rangle \mathrm{d}V$ nor the total magnetic fluctuational helicity $\int_V \langle \mathbf{a}' \cdot \mathbf{b}' \rangle \mathrm{d}V$ need to be conserved during the evolution of the flow and only their sum is an evolutional invariant. By the use of (2.28b) and (2.43) the evolution of the fluctuational magnetic helicity is governed by the following equation

$$\begin{aligned}\frac{\mathrm{D}}{\mathrm{D}t}\langle \mathbf{a}' \cdot \mathbf{b}' \rangle &= -2\boldsymbol{\mathcal{E}} \cdot \langle \mathbf{B} \rangle \\ &\quad + \nabla \cdot [\langle \mathbf{u}' \cdot \mathbf{a}' \rangle \langle \mathbf{B} \rangle + \langle [(\langle \mathbf{u} \rangle + \mathbf{u}') \cdot \mathbf{a}' - \phi']\mathbf{b}' \rangle - \langle (\mathbf{a}' \cdot \langle \mathbf{B} \rangle)\mathbf{u}' \rangle] \\ &\quad - \frac{\partial}{\partial x_m}\left[\left\langle u_m' a_j' b_j' \right\rangle - u_m'\left\langle a_j' b_j' \right\rangle\right] + \eta\langle \nabla^2\mathbf{a}' \cdot \mathbf{b}' \rangle + \eta\langle \mathbf{a}' \cdot \nabla^2\mathbf{b}' \rangle.\end{aligned} \tag{2.49}$$

If the gauge is chosen such that $\nabla \cdot \mathbf{a}' = 0$ the last diffusive terms can be rearranged

$$\langle \nabla^2\mathbf{a}' \cdot \mathbf{b}' \rangle = -\langle \nabla \times \mathbf{b}' \cdot \mathbf{b}' \rangle = -\langle \mathbf{j}' \cdot \mathbf{b}' \rangle, \tag{2.50a}$$

$$\langle \mathbf{a}' \cdot \nabla^2\mathbf{b}' \rangle = -\langle \mathbf{a}' \cdot \nabla \times \mathbf{j}' \rangle = -\left\langle \epsilon_{ijk} a_i' \partial_j j_k' \right\rangle = -\langle \mathbf{j}' \cdot \mathbf{b}' \rangle - \epsilon_{ijk}\left\langle \partial_j\left(j_k' a_i'\right)\right\rangle, \tag{2.50b}$$

to yield

$$\eta\langle \nabla^2\mathbf{a}' \cdot \mathbf{b}' \rangle + \eta\langle \mathbf{a}' \cdot \nabla^2\mathbf{b}' \rangle = -2\eta\langle \mathbf{j}' \cdot \mathbf{b}' \rangle + \eta\nabla \cdot \langle \mathbf{a}' \times \mathbf{j}' \rangle. \tag{2.51}$$

We observe that the term $\boldsymbol{\mathcal{E}} \cdot \langle \mathbf{B} \rangle$ is an important production mechanism of the fluctuational magnetic helicity.

2.1.1.2 Cross-helicity

The mean cross-helicity is also separated into the cross-helicity of the mean field and the fluctuational component

$$\langle \mathbf{U} \cdot \mathbf{B} \rangle = \langle \mathbf{U} \rangle \cdot \langle \mathbf{B} \rangle + \langle \mathbf{u}' \cdot \mathbf{b}' \rangle. \tag{2.52}$$

By (2.6a and b) the total cross-helicity $\mathcal{H}_c = \int_V \mathbf{U} \cdot \mathbf{B} \mathrm{d}V$ in the entire fluid volume V satisfies

$$\begin{aligned} \frac{\mathrm{D}}{\mathrm{D}t} \int_V \mathbf{U} \cdot \mathbf{B} \mathrm{d}V = &- \int_V (\mathbf{U} \times \mathbf{B}) \cdot \mathbf{\Omega} \mathrm{d}V - \int_V \mathbf{f} \cdot \mathbf{B} \mathrm{d}V - (\nu + \eta) \int_V \mathbf{W} \cdot \mathbf{J} \mathrm{d}V \\ &+ \int_V \nabla \cdot \left[\nu \mathbf{B} \times \mathbf{W} + \eta \mathbf{U} \times \mathbf{J} - \Pi \mathbf{B} + \mathbf{B} \left(\frac{U^2 + B^2}{2} \right) \right] \mathrm{d}V \\ = &- \int_V (\mathbf{U} \times \mathbf{B}) \cdot \mathbf{\Omega} \mathrm{d}V - \int_V \mathbf{f} \cdot \mathbf{B} \mathrm{d}V - (\nu + \eta) \int_V \mathbf{W} \cdot \mathbf{J} \mathrm{d}V, \end{aligned} \tag{2.53}$$

and the last equality is obtained under the assumption that the flux of the magnetic field through the boundary of the domain vanishes and so do the velocity and $\mathbf{B} \times \mathbf{W}$ on the boundary. Hence the total cross-helicity is conserved in the absence of viscosity, $\nu = 0$, magnetic diffusion $\eta = 0$, the Coriolis force, $\mathbf{\Omega} = 0$, and the external forcing $\mathbf{f} = 0$. When $\nu = \eta = 0$, $\mathbf{\Omega} = 0$, and $\mathbf{f} = 0$ we have

$$\int_V \langle \mathbf{U} \rangle \cdot \langle \mathbf{B} \rangle \mathrm{d}V = \mathcal{H}_c - \int_V \langle \mathbf{u}' \cdot \mathbf{b}' \rangle \mathrm{d}V. \tag{2.54}$$

The evolution equation for the fluctuational cross-helicity can be cast in the form (cf (2.28a and b))

$$\begin{aligned} \frac{\mathrm{D}}{\mathrm{D}t} \langle \mathbf{u}' \cdot \mathbf{b}' \rangle = &- \boldsymbol{\mathcal{E}} \cdot (\langle \mathbf{W} \rangle + 2\mathbf{\Omega}) - \left\langle u_i' u_j' - b_i' b_j' \right\rangle \frac{\partial \langle B \rangle_i}{\partial x_j} + \langle \mathbf{f} \cdot \mathbf{b}' \rangle \\ &+ \nabla \cdot \left[\left\langle \left(-\Pi' + \frac{u'^2 + b'^2}{2} \right) \mathbf{b}' \right\rangle + \left\langle \frac{u'^2 + b'^2}{2} \right\rangle \langle \mathbf{B} \rangle \right. \\ &\left. - \nu \langle \mathbf{w}' \times \mathbf{b}' \rangle + \eta \langle \mathbf{u}' \times \mathbf{j}' \rangle \right] \\ &- \frac{\partial}{\partial x_m} \left[\left\langle u_m' u_j' b_j' \right\rangle - u_m' \left\langle u_j' b_j' \right\rangle \right] - (\nu + \eta) \langle \mathbf{w}' \cdot \mathbf{j}' \rangle. \end{aligned} \tag{2.55}$$

Of course the EMF $\boldsymbol{\mathcal{E}}$ is given by (2.26) and the effect $-\boldsymbol{\beta} \cdot \langle \mathbf{J} \rangle \cdot (\langle \mathbf{W} \rangle + 2\mathbf{\Omega})$ has been argued by [12] to be one of the major mechanisms of cross-helicity production; similarly, the flux term $\nabla \cdot [\langle u'^2 + b'^2 \rangle \langle \mathbf{B} \rangle / 2]$. In addition, as shown by [23], another important cross-helicity production mechanism is associated with the term involving the Reynolds–Maxwell stresses $\left\langle u_i' u_j' - b_i' b_j' \right\rangle$ and the mean-field gradient; see also section 5.3 below.

2.1.1.3 Kinetic helicity

Next we turn to the mean kinetic helicity,

$$\langle \mathbf{U} \cdot \mathbf{W} \rangle = \langle \mathbf{U} \rangle \cdot \langle \mathbf{W} \rangle + \langle \mathbf{u}' \cdot \mathbf{w}' \rangle. \tag{2.56}$$

The total kinetic helicity $\mathcal{H}_k = \int_V \mathbf{U} \cdot \mathbf{W} \mathrm{d}V$ in the entire fluid volume V satisfies (cf (2.6a) and (2.42b)),

$$\begin{aligned}
\frac{\mathrm{D}}{\mathrm{D}t} \int_V \mathbf{U} \cdot \mathbf{W} \mathrm{d}V &= 2 \int_V [(\mathbf{f} - 2\boldsymbol{\Omega} \times \mathbf{U} + \mathbf{J} \times \mathbf{B}) \cdot \mathbf{W} - \nu \mathbf{W} \cdot \nabla \times \mathbf{W}] \mathrm{d}V \\
&\quad + \int_V \nabla \cdot \left[\mathbf{f} \times \mathbf{U} + \left(-\Pi + \frac{B^2}{2} + \frac{U^2}{2}\right) \mathbf{W} + \nu \mathbf{U} \times (\nabla \times \mathbf{W})\right. \\
&\quad \left. + 2\boldsymbol{\Omega} U^2 - 2\mathbf{U}(\boldsymbol{\Omega} \cdot \mathbf{U}) + \mathbf{B}(\mathbf{U} \cdot \mathbf{J}) - \mathbf{J}(\mathbf{U} \cdot \mathbf{B})\right] \mathrm{d}V \\
&= 2 \int_V [(\mathbf{f} - 2\boldsymbol{\Omega} \times \mathbf{U} + \mathbf{J} \times \mathbf{B}) \cdot \mathbf{W} - \nu \mathbf{W} \cdot \nabla \times \mathbf{W}] \mathrm{d}V,
\end{aligned} \tag{2.57}$$

where the last equality is obtained under the assumption that the total kinetic helicity flux vanishes on the boundary of the domain. In the absence of forcing, $\mathbf{f} = 0$, background rotation $\boldsymbol{\Omega} = 0$, the magnetic field $\mathbf{B} = 0$ and viscosity $\nu = 0$ the total kinetic helicity $\mathcal{H}_k$ is conserved and

$$\int_V \langle \mathbf{U} \rangle \cdot \langle \mathbf{W} \rangle \mathrm{d}V = \mathcal{H}_k - \int_V \langle \mathbf{u}' \cdot \mathbf{w}' \rangle \mathrm{d}V. \tag{2.58}$$

By the use of (2.28a) and (2.44) we can obtain the evolution equation of the fluctuational kinetic helicity

$$\begin{aligned}
\frac{\mathrm{D}}{\mathrm{D}t} \langle \mathbf{u}' \cdot \mathbf{w}' \rangle &= -2 \langle u_i' u_j' \rangle \frac{\partial \langle W \rangle_i}{\partial x_j} + 2 \langle \mathbf{w}' \times \mathbf{j}' \rangle \cdot \langle \mathbf{B} \rangle - 2 \langle \mathbf{w}' \times \mathbf{b}' \rangle \cdot \langle \mathbf{J} \rangle \\
&\quad - 2 \left\langle b_i' b_j' \frac{\partial w_i'}{\partial x_j} \right\rangle - 2\nu \langle \mathbf{w}' \cdot \nabla \times \mathbf{w}' \rangle + 2 \langle \mathbf{f} \cdot \mathbf{w}' \rangle \\
&\quad + \nabla \cdot \left[\langle \mathbf{f} \times \mathbf{u}' \rangle - \langle \Pi' \mathbf{w}' \rangle + \left\langle \mathbf{w}' \frac{u'^2 - b'^2}{2} \right\rangle + 2 \langle \mathbf{b}' \mathbf{b}' \cdot \mathbf{w}' \rangle \right. \\
&\quad + \langle \mathbf{b}' \mathbf{u}' \cdot \mathbf{j}' \rangle - \langle \mathbf{j}' \mathbf{u}' \cdot \mathbf{b}' \rangle + (\langle \mathbf{W} \rangle + 2\boldsymbol{\Omega}) \cdot \langle \mathbf{u}' \mathbf{u}' \rangle \\
&\quad + \langle \mathbf{w}' b_j' - \mathbf{j}' u_j' \rangle \langle B \rangle_j + \langle \mathbf{u}' \cdot \mathbf{j}' \rangle \langle \mathbf{B} \rangle + \langle \mathbf{b}' u_j' \rangle \cdot \langle J \rangle_j \\
&\quad \left. - \langle \mathbf{u}' \cdot \mathbf{b}' \rangle \langle \mathbf{J} \rangle + \nu \left\langle \mathbf{u}' \times \left(\nabla \times \mathbf{w}' \right) \right\rangle \right].
\end{aligned} \tag{2.59}$$

2.1.1.4 Current helicity

The α-effect in stationary, homogeneous and isotropic turbulence is proportional to the full residual helicity $\langle h_r \rangle = -\langle \mathbf{u}' \cdot \mathbf{w}' \rangle + \langle \mathbf{b}' \cdot \mathbf{j}' \rangle$, as in (2.40). Therefore, in order to complete the dynamical picture of the EMF evolution we also need the mean current helicity,

$$\langle \mathbf{B} \cdot \mathbf{J} \rangle = \langle \mathbf{B} \rangle \cdot \langle \mathbf{J} \rangle + \langle \mathbf{b}' \cdot \mathbf{j}' \rangle. \tag{2.60}$$

From (2.6b) and (2.42c) the total current helicity $\mathcal{H}_{cur} = \int_V \mathbf{B} \cdot \mathbf{J} \mathrm{d}V$ in the entire fluid volume V satisfies

$$\begin{aligned}\frac{\mathrm{D}}{\mathrm{D}t}\int_V \mathbf{B}\cdot\mathbf{J}\mathrm{d}V &= 2\int_V [(\mathbf{U}\times\mathbf{B})\cdot\nabla\times\mathbf{J} - \eta\mathbf{J}\cdot\nabla\times\mathbf{J}]\mathrm{d}V + \\ &\int_V \nabla\cdot[2\mathbf{B}\mathbf{U}\cdot\mathbf{J} - \mathbf{U}\mathbf{B}\cdot\mathbf{J} - \mathbf{B}\times\nabla\times(\mathbf{U}\times\mathbf{B}) + \eta\mathbf{B}\times\nabla\times\mathbf{J}] \\ &= 2\int_V [(\mathbf{U}\times\mathbf{B})\cdot\nabla\times\mathbf{J} - \eta\mathbf{J}\cdot\nabla\times\mathbf{J}]\mathrm{d}V.\end{aligned} \tag{2.61}$$

Similarly as before, the last equality is obtained under the assumption that the total current helicity flux vanishes on the boundary of the domain. By (2.28b) and (2.45) the fluctuational current helicity evolves according to

$$\begin{aligned}\frac{\mathrm{D}}{\mathrm{D}t}\langle\mathbf{b}'\cdot\mathbf{j}'\rangle &= 2\langle\mathbf{b}'\times\nabla\times\mathbf{j}'\rangle\cdot\langle\mathbf{U}\rangle - 2\langle\mathbf{u}'\times\nabla\times\mathbf{j}'\rangle\cdot\langle\mathbf{B}\rangle \\ &+ 2\langle(\mathbf{u}'\times\mathbf{b}')\cdot\nabla\times\mathbf{j}'\rangle - 2\eta\langle\mathbf{j}'\cdot\nabla\times\mathbf{j}'\rangle \\ &+ \nabla\cdot\Big[2\langle\mathbf{u}'\cdot\mathbf{j}'\rangle\langle\mathbf{B}\rangle - \langle\mathbf{b}'\cdot\mathbf{j}'\rangle\langle\mathbf{U}\rangle + 2\langle\mathbf{b}'j_i'\rangle\langle U\rangle_i - 2\langle\mathbf{u}'j_i'\rangle\langle B\rangle_i \\ &\quad - \langle\mathbf{b}'\times\nabla\times(\mathbf{u}'\times\langle\mathbf{B}\rangle + \langle\mathbf{U}\rangle\times\mathbf{b}')\rangle + 2\langle\mathbf{b}'\mathbf{u}'\cdot\mathbf{j}'\rangle \\ &\quad - \langle\mathbf{u}'\mathbf{j}'\cdot\mathbf{b}'\rangle - \langle\mathbf{b}'\times\nabla\times(\mathbf{u}'\times\mathbf{b}')\rangle + \eta\langle\mathbf{b}'\times(\nabla\times\mathbf{j}')\rangle\Big].\end{aligned} \tag{2.62}$$

Since the residual helicity is a sum of the kinetic helicity (with a minus sign) and the current helicity, the evolution equation for the residual helicity is readily obtained from (2.59) and (2.62). Although under the so-called *First-Order Smoothing Approximation* (FOSA), which corresponds to the weak turbulence regime (see [2] and chapter 3 for definition of FOSA) and in the weak seed field limit $\| \langle\mathbf{B}\rangle \| \ll \| \mathbf{U} \|$ the kinetic and current helicities are related, in a general, fully nonlinear case these two quantities are largely independent. For example, a radial and radially dependent flow of compressible plasma in a cylindrical cavity induced by a relaxation process of a poloidal, radially dependent magnetic field (as considered by [24]) would possess non-zero current helicity, but no kinetic helicity. On the other hand the kinetic helicity can be obviously non-zero in a flow even in the absence of magnetic fields.

2.2 Effects of non-equilibrium turbulence in formation of the EMF—simplified picture

The spatially averaged mean electromotive force generated by turbulent fluctuations takes the form

$$\begin{aligned}\boldsymbol{\mathcal{E}} &= \langle\mathbf{u}'\times\mathbf{b}'\rangle_{\mathbf{x}} \\ &= \left\langle\int \mathrm{d}^3k\int\mathrm{d}\omega\int\mathrm{d}^3k'\int\mathrm{d}\omega'\hat{u}_j(\mathbf{k},\omega)\hat{b}_k(\mathbf{k}',\omega')\mathrm{e}^{\mathrm{i}[(\mathbf{k}+\mathbf{k}')\cdot\mathbf{x}-(\omega+\omega')t]}\right\rangle_{\mathbf{x}}\epsilon_{ijk}\hat{e}_k \\ &= \int\mathrm{d}^3k\int\mathrm{d}\omega\int\mathrm{d}\omega'\hat{u}_j(\mathbf{k},\omega)\hat{b}_k^*(\mathbf{k},\omega')\mathrm{e}^{-\mathrm{i}(\omega-\omega')t},\end{aligned} \tag{2.63}$$

where

$$\langle\cdot\rangle_{\mathbf{x}} = \frac{3}{4\pi L^3}\int_{|\mathbf{r}|<L} \psi(\mathbf{x}+\mathbf{r},\, t)\mathrm{d}^3r, \tag{2.64}$$

denotes a spatial average over any large-scale $L \gg \ell_e$. The typical assumption at this stage concerns the statistical properties of turbulence and typically homogeneity and stationarity had been assumed. In case of real MHD turbulence both of these assumptions should be relaxed, however, here we will concentrate mainly on the effect of non-stationarity. The above formula for the mean EMF greatly simplifies in the case of only two Fourier modes in time, ω_1, ω_2 and one in space, $\mathbf{k}$

$$\begin{aligned}\boldsymbol{\mathcal{E}}(\mathbf{k}) &= \frac{1}{2}\,\mathfrak{Re}\left[(\hat{\mathbf{u}}^{(1)}\mathrm{e}^{\mathrm{i}\omega_1 t} + \hat{\mathbf{u}}^{(2)}\mathrm{e}^{\mathrm{i}\omega_2 t})\mathrm{e}^{\mathrm{i}\mathbf{k}\cdot\mathbf{x}} \times \left(\hat{\mathbf{b}}^{(1)*}\mathrm{e}^{-\mathrm{i}\omega_1 t} + \hat{\mathbf{b}}^{(2)*}\mathrm{e}^{-\mathrm{i}\omega_2 t}\right)\mathrm{e}^{-\mathrm{i}\mathbf{k}\cdot\mathbf{x}}\right]\\ &= \frac{1}{2}\,\mathfrak{Re}\left[\hat{\mathbf{u}}^{(1)} \times \hat{\mathbf{b}}^{(2)*}\mathrm{e}^{\mathrm{i}\Delta\omega t} + \hat{\mathbf{u}}^{(2)} \times \hat{\mathbf{b}}^{(1)*}\mathrm{e}^{-\mathrm{i}\Delta\omega t}\right]\\ &\quad+ \frac{1}{2}\,\mathfrak{Re}\left[\hat{\mathbf{u}}^{(1)} \times \hat{\mathbf{b}}^{(1)*} + \hat{\mathbf{u}}^{(2)} \times \hat{\mathbf{b}}^{(2)*}\right],\end{aligned} \tag{2.65}$$

where $\Delta\omega = \omega_1 - \omega_2$, $\mathfrak{Re}$ denotes the real part and star the complex conjugate. Within the common approach of statistically stationary turbulence the first term in square brackets in the above formula (2.65) is absent and the second term in square brackets can only be non-zero if the magnetic diffusion is present, which creates a phase shift between the velocity and magnetic fields; the latter is a well-known result thoroughly explained e.g. in [2, 25]. Without the magnetic diffusion the vectorial multiplications between the velocity and magnetic fields belonging to the same spatial and temporal Fourier modes are purely imaginary and hence the large-scale EMF is zero.

However, in general the term

$$\boldsymbol{\mathcal{E}}_{\mathrm{neq}}(\mathbf{k}) = \frac{1}{2}\,\mathfrak{Re}\left[\hat{\mathbf{u}}^{(1)} \times \hat{\mathbf{b}}^{(2)*}\mathrm{e}^{\mathrm{i}\Delta\omega t} + \hat{\mathbf{u}}^{(2)} \times \hat{\mathbf{b}}^{(1)*}\mathrm{e}^{-\mathrm{i}\Delta\omega t}\right], \tag{2.66}$$

is non-zero and its influence on the large-scale dynamos will be studied here. It needs to be stressed, that the latter contribution to the large-scale EMF which results from non-equilibrium effects depends on time on the scale $2\pi/\Delta\omega$, i.e. determined by the difference between the frequencies of two Fourier modes. When $\Delta\omega$ is very small, the time-scale of evolution of the mean EMF can be much larger than the time-scale of variation of the turbulent fluctuations and even much larger than the typical variation time-scale of the mean field. Then the time average of $\boldsymbol{\mathcal{E}}$ can be non-zero and such an EMF can lead to growth and equilibration of the mean magnetic field. Another possibility for non-vanishing mean EMF in (2.66) is when at least one of the interacting modes is unstable, i.e. ω is complex and $\exp(\mathrm{i}\Delta\omega t) \sim \exp[-\mathfrak{Im}(\Delta\omega)t]$, with $\mathfrak{Im}\,(\Delta\omega) < 0$. However, the latter case requires an underlying MHD instability in the flow and it will not be pursued in here[2].

References

[1] Roberts P H 1967 *An Introduction to Magnetohydrodynamics* (London: Longmans Green)

[2] Moffatt H K and Dormy E 2019 *Self-exciting Fluid Dynamos* (Cambridge: Cambridge University Press)

[3] Steenbeck M, Krause F and Rädler K-H 1966 A calculation of the mean electromotive force in an electrically conducting fluid in turbulent motion, under the influence of Coriolis forces *Z. Naturforsch.* **21** 369–76 [English translation: Roberts & Stix (1971) pp 29–47]

[4] Hughes D W and Tobias S M 2010 An introduction to mean field dynamo theory *Reviews of the Theory of Magnetized Plasmas, Volume 1: Relaxation Dynamics in Laboratory and Astrophysical Plasmas* ed P H Diamond, X Garbet, P Ghendrih and Y Sarazin (Singapore: World Scientific) pp 15–48

[5] Childress S and Soward A M 1972 Convection-driven hydromagnetic dynamo *Phys. Rev. Lett.* **29** 837–9

[6] Soward A M 1974 A convection-driven dynamo I. The weak field case *Phil. Trans. R. Soc. Lond.* A **275** 611–51

[7] Mizerski K A and Tobias S M 2013 Large-scale convective dynamos in a stratified rotating plane layer *Geophys. Astrophys. Fluid Dyn.* **107** 218–43

[8] Calkins M A, Julien K, Tobias S M and Aurnou J M 2015 A multiscale dynamo model driven by quasi-geostrophic convection *J. Fluid Mech.* **780** 143–66

[9] Calkins M A 2018 Quasi-geostrophic dynamo theory *Phys. Earth Planet. Inter.* **276** 182–9

[10] Pouquet A, Frisch U and Léorat J 1976 Strong MHD helical turbulence and the nonlinear dynamo effect *J. Fluid Mech.* **77** 321–54

[11] Yoshizawa A 1998 *Hydrodynamic and Magnetohydrodynamic Turbulent Flows: Modelling and Statistical Theory* (Dordrecht: Springer)

[12] Yokoi N 2013 Cross helicity and related dynamo *Geophys. Astrophys. Fluid Dyn.* **107** 114–84

[13] Mizerski K A, Yokoi N and Brandenburg A 2023 Cross-helicity effect on α-type dynamo in non-equilibrium turbulence *J. Plasma Phys.* **89** 905890412

[14] Yoshizawa A 1990 Self-consistent turbulent dynamo modeling of reversed field pinches and planetary magnetic fields *Phys. Fluids* B **2** 1589–600

[15] Sur S and Brandenburg A 2009 The role of the Yoshizawa effect in the Archontis dynamo *Mon. Not. R. Astron. Soc.* **399** 273–80

---

[2] An interesting situation could hypothetically take place in a system with multiple and non-degenerate eigenvalue, i.e. associated with multiple eigenmodes of the linearized equations (perhaps present due to some type of forcing). In such a case the oscillations would not be present but instead the mean EMF would grow in time as t^n, where n is a natural number dependent on the multiplicity of the eigenvalue; for a double, non-degenerate eigenvalue in the absence of magnetic diffusion one would have

$$\begin{aligned}\boldsymbol{\mathcal{E}}(\mathbf{k}) &= \frac{1}{2}\,\mathfrak{Re}\left[\left(\hat{\mathbf{u}}^{(1)} + t\hat{\mathbf{u}}^{(2)}\right)e^{i(\mathbf{k}\cdot\mathbf{x}+\omega t)} \times \left(\hat{\mathbf{b}}^{(1)*} + t\hat{\mathbf{b}}^{(2)*}\right)e^{-i(\mathbf{k}\cdot\mathbf{x}+\omega t)}\right] \\ &= \frac{1}{2}t\,\mathfrak{Re}\left[\hat{\mathbf{u}}^{(1)} \times \hat{\mathbf{b}}^{(2)*} + \hat{\mathbf{u}}^{(2)} \times \hat{\mathbf{b}}^{(1)*}\right].\end{aligned} \tag{2.67}$$

[16] Yokoi N 2023 Unappreciated cross-helicity effects in plasma physics: anti-diffusion effects in dynamo and momentum transport *Rev. Mod. Plasma Phys.* **7** 33

[17] Kraichnan R H 1959 The structure of isotropic turbulence at very high Reynolds numbers *J. Fluid Mech.* **5** 497–543

[18] Yokoi N 2018 Electromotive force in strongly compressible magnetohydrodynamic turbulence *J. Plasma Phys.* **84** 735840501l

[19] Rädler K-H 1969 A new turbulent dynamo. I *Monats. Dt. Akad. Wiss.* **11** 272–9 [English translation: Roberts P H and Stix M 1971 The turbulent dynamo: a translation of a series of papers by F Krause, K-H Rädler and M Steenbeck *NCAR Technical Note NCAR/TN-60 +IA* pp 301–8] http://dx.doi.org/10.5065/D6DJ5CK7

[20] Rogachevskii I and Kleeorin N 2003 Electromotive force and large-scale magnetic dynamo in a turbulent flow with a mean shear *Phys. Rev.* E **68** 036301

[21] Rogachevskii I and Kleeorin N 2004 Nonlinear theory of a 'shear-current' effect and mean-field magnetic dynamos *Phys. Rev.* E **70** 046310

[22] Rogachevskii I, Kleeorin N and Liverts E 2006 Nonlinear shear-current dynamo and magnetic helicity transport in sheared turbulence *Geophys. Astrophys. Fluid Dyn.* **100** 537–57

[23] Yokoi N and Hoshino M 2011 Flow-turbulence interaction in magnetic reconnection *Phys. Plasmas* **18** 111208

[24] Moffatt H K and Mizerski K A 2018 Pinch dynamics in a low-β plasma *Fluid Dyn. Res.* **50** 011401

[25] Moffatt H K and Proctor M R E 1982 The role of the helicity spectrum function in turbulent dynamo theory *Geophys. Astrophys. Fluid Dyn.* **21** 265–83

Chapter 3

Weak turbulence: effect of beating waves

The aim of this chapter is to demonstrate the effect of interactions between distinct waves induced, e.g. by an oscillatory forcing, on generation of the large-scale electromotive force (EMF) in highly electrically conducting plasma in the dynamical regime, i.e. when the back reaction of the Lorentz force on the flow is significant. The mean EMF is induced by waves forced at resonance (or close to resonance) with the system eigen frequencies, corresponding to the frequencies of the well-known 'Lehnert' (Magneto-Coriolis, MC) waves. These fundamental magnetohydrodynamic (MHD) wave modes which are essentially Alfvén waves modified by the effect of the Coriolis force, were first studied by Lehnert (1954) [1] and therefore will be named here the 'Lehnert waves'. The formulae for the mean EMF both for an ideal and highly conducting plasma are derived here, within the framework of the theory of weak, unbalanced MHD turbulence (with Coriolis force), which is essentially linear in the turbulent fluctuations and unbalanced so that the up-down symmetry is broken and the net helicity of turbulent flow has a preferred sign (the Lehnert waves may have either positive or negative helicity, and it is supposed that waves of a single sign of helicity are preferentially excited by a symmetry-breaking mechanism, cf [2, 3]). The regime of weak turbulence corresponds to weak amplitude of turbulent pulsations and linearization of their evolution; this idea has been thoroughly explained and related to the general context of fully developed turbulence and MHD turbulence in [4, 5] and [6]. Although the Lehnert wave packets interact nonlinearly, when their amplitudes are small, they can often survive for a long time without getting destroyed, which is the essential feature of the weak MHD turbulence, as also explained in [3]. In particular, Monsalve *et al* [7] report quantitative experimental observation of a persistent weak regime in a three-dimensional inertial-wave turbulence in a rotating system. Nevertheless, the theory of weak turbulence predicts, that it must eventually evolve into the strongly nonlinear regime; in other words even in the situations, when the weak turbulence can persist, at large times the terms nonlinear in the fluctuations start to play a significant role in their dynamics. An interested reader is directed to [8], for a

doi:10.1088/978-0-7503-6362-4ch3

review of various approaches and a physical description of turbulent flows, in particular chapter 10 of that book, concerned with buoyancy driven turbulence.

Consequently, in this chapter we consider linearized evolution of the turbulent fluctuations, i.e. the nonlinear terms in the equations (2.13a–c) are neglected, which yields

$$\begin{aligned}\frac{\partial \mathbf{u}'}{\partial t} - \nu\nabla^2\mathbf{u}' + 2\boldsymbol{\Omega}\times\mathbf{u}' + (\langle\mathbf{U}\rangle\cdot\nabla)\mathbf{u}' - (\langle\mathbf{B}\rangle\cdot\nabla)\mathbf{b}' + \nabla\Pi' \\ = \mathbf{f} - (\mathbf{u}'\cdot\nabla)\langle\mathbf{U}\rangle + (\mathbf{b}'\cdot\nabla)\langle\mathbf{B}\rangle,\end{aligned} \tag{3.1a}$$

$$\frac{\partial \mathbf{b}'}{\partial t} - \eta\nabla^2\mathbf{b}' + (\langle\mathbf{U}\rangle\cdot\nabla)\mathbf{b}' - (\langle\mathbf{B}\rangle\cdot\nabla)\mathbf{u}' = -(\mathbf{u}'\cdot\nabla)\langle\mathbf{B}\rangle + (\mathbf{b}'\cdot\nabla)\langle\mathbf{U}\rangle, \tag{3.1b}$$

$$\nabla\cdot\mathbf{b}' = 0, \qquad \nabla\cdot\mathbf{u}' = 0. \tag{3.1c}$$

The force field $\mathbf{f}$ is assumed sufficiently weak, so that the fluctuations can remain small and the linearization remains valid.

3.1 Lehnert waves

In this chapter on weak turbulence, when the evolution of the fluctuations is linear, we can concentrate on a single spatial Fourier mode and, therefore, it is more convenient to use the spatial average (2.64) over a large scale, $L \gg \ell_e$. This allows us to calculate mean wave interactions of Fourier modes

$$\mathbf{u}' = \mathfrak{Re}\,\hat{\mathbf{u}}(t)e^{i\mathbf{k}\cdot\mathbf{x}}, \qquad \mathbf{b}' = \mathfrak{Re}\,\hat{\mathbf{b}}(t)e^{i\mathbf{k}\cdot\mathbf{x}}, \qquad \Pi' = \mathfrak{Re}\,\hat{\Pi}(t)e^{i\mathbf{k}\cdot\mathbf{x}}, \tag{3.2}$$

in a simple way, namely

$$\left\langle u'_i b'_j \right\rangle_{\mathbf{x}} = \frac{1}{2}\mathfrak{Re}\left(\hat{u}'_i \hat{b}'^*_j\right), \tag{3.3}$$

where $\mathfrak{Re}$ denotes the real part and the star denotes a complex conjugate. In addition, for the purpose of this section and the following one, we neglect the effect of large-scale gradients in the fluctuational equations (3.1a and b), in other words we assume that the scale separation is significant enough for the gradients of the large-scale fields to be negligibly weak in the dynamics of fluctuations; consequently, we consider the following system of fluctuational equations

$$\frac{\partial \mathbf{u}'}{\partial t} - \nu\nabla^2\mathbf{u}' + 2\boldsymbol{\Omega}\times\mathbf{u}' + (\langle\mathbf{U}\rangle\cdot\nabla)\mathbf{u}' - (\langle\mathbf{B}\rangle\cdot\nabla)\mathbf{b}' + \nabla\Pi' = \mathbf{f}, \tag{3.4a}$$

$$\frac{\partial \mathbf{b}'}{\partial t} - \eta\nabla^2\mathbf{b}' + (\langle\mathbf{U}\rangle\cdot\nabla)\mathbf{b}' - (\langle\mathbf{B}\rangle\cdot\nabla)\mathbf{u}' = 0, \tag{3.4b}$$

$$\nabla\cdot\mathbf{b}' = 0, \qquad \nabla\cdot\mathbf{u}' = 0. \tag{3.4c}$$

We will now find the eigen modes of the linearized MHD equations (3.4a–c) in the absence of diffusion, $\nu = \eta = 0$, known as the Lehnert waves (or the MC waves), and calculate the mean EMF generated from interactions of two such waves with distinct eigen frequencies. Since the resistivity is neglected we eliminate the effect of the '*self-mode*' interactions of the type $\mathfrak{Re}\left[\hat{\mathbf{u}}^{(i)} \times \hat{\mathbf{b}}^{(i)*}\right] = 0$ on the large-scale EMF, as explained below (2.65).

In order to find the dispersion relation we must also consider time Fourier modes, i.e. introduce explicit spectral frequency of time oscillations

$$\mathbf{u}' = \mathfrak{Re}\, \hat{\mathbf{u}} e^{i(\mathbf{k}\cdot\mathbf{x}-\omega t)}, \qquad \mathbf{b}' = \mathfrak{Re}\, \hat{\mathbf{b}} e^{i(\mathbf{k}\cdot\mathbf{x}-\omega t)}, \qquad \Pi' = \mathfrak{Re}\, \hat{\Pi} e^{i(\mathbf{k}\cdot\mathbf{x}-\omega t)}, \tag{3.5}$$

and under negligible dissipation, $\nu \approx \eta \approx 0$, the linearized equations (3.4a–c) take the form

$$-i\omega\hat{\mathbf{u}} = -i(\langle\mathbf{U}\rangle\cdot\mathbf{k})\hat{\mathbf{u}} - i\mathbf{k}\hat{\Pi} - 2\Omega\hat{\mathbf{e}}_z \times \hat{\mathbf{u}} + i(\langle\mathbf{B}\rangle\cdot\mathbf{k})\hat{\mathbf{b}}, \tag{3.6a}$$

$$-i\omega\hat{\mathbf{b}} = -i(\langle\mathbf{U}\rangle\cdot\mathbf{k})\hat{\mathbf{b}} + i(\langle\mathbf{B}\rangle\cdot\mathbf{k})\hat{\mathbf{u}}, \tag{3.6b}$$

$$\mathbf{k}\cdot\hat{\mathbf{u}} = 0 \qquad \mathbf{k}\cdot\hat{\mathbf{b}} = 0. \tag{3.6c}$$

We now observe that the effect of the mean flow $\langle\mathbf{U}\rangle$ is easily incorporated into a modified frequency $\omega \to \omega + i\langle\mathbf{U}\rangle\cdot\mathbf{k}$, that is, by local Galilean transformation to a frame moving with velocity $\langle\mathbf{U}\rangle$; hence in the following the terms involving $\langle\mathbf{U}\rangle\cdot\mathbf{k}$ in (3.6a and b) will be disregarded. Contracting the Navier–Stokes equation with $\mathbf{k}$ (acting on both sides with $\mathbf{k}\cdot$) one obtains $\hat{\Pi} = -i(2\Omega/k^2)\hat{\mathbf{e}}_z\cdot(\mathbf{k}\times\hat{\mathbf{u}})$, which can be introduced back into (3.6). Next, noting that the vertical z components of the fields separate from the horizontal components x and y (thus vertical components of $\hat{\mathbf{u}}$ and $\hat{\mathbf{b}}$ can be calculated from the horizontal ones), we obtain the following equation for the eigenvalue ω,

$$\begin{vmatrix} 2\Omega\dfrac{k_xk_y}{k^2} + i\omega & 2\Omega\dfrac{k_y^2 + k_z^2}{k^2} & i\langle\mathbf{B}\rangle\cdot\mathbf{k} & 0 \\ -2\Omega\dfrac{k_x^2 + k_z^2}{k^2} & -2\Omega\dfrac{k_xk_y}{k^2} + i\omega & 0 & i\langle\mathbf{B}\rangle\cdot\mathbf{k} \\ i\langle\mathbf{B}\rangle\cdot\mathbf{k} & 0 & i\omega & 0 \\ 0 & i\langle\mathbf{B}\rangle\cdot\mathbf{k} & 0 & i\omega \end{vmatrix} = 0. \tag{3.7}$$

The solution for the eigen values takes the form

$$\omega \in \left[\pm\Omega\frac{k_z}{k} \pm_u \sqrt{\Omega^2\frac{k_z^2}{k^2} + (\langle\mathbf{B}\rangle\cdot\mathbf{k})^2}\right], \tag{3.8}$$

where the + and − signs in $\pm_u$ are uncorrelated with the preceding ± sign in (3.8). This is a dispersion relation of the MC waves, which have been known since Lehnert (1954) [1] as some of the fundamental MHD modes (for a review see e.g. [9]). The linearized system (3.6a–c) admits solutions in the form of Lehnert waves, which in the absence of dissipation are neither damped nor amplified and according to the weak turbulence picture can sometimes survive a long time in natural systems (see [3]). However, many numerical simulations, such as e.g. in the problem of rapidly rotating turbulent convection studied by [10, 11] indicate, that when diffusion is very weak compared to advection and straining, the amplitude of fluctuations can be amplified quickly, effectively destroying the linear regime. We emphasize, therefore, that the assumption of weak turbulence is typically not valid, in particular, for small-scale systems, and turbulence develops into a strongly nonlinear regime studied here in later chapters. Nevertheless, Hughes and Cattaneo (2008) [11] conclude, that at low diffusivity the size of the system plays a crucial role for generation of large-scale fields and the EMFs obtained in their simulations strongly oscillate in time. The latter result is in direct correspondence to the results reported here for both weak and strong non-equilibrium turbulence.

We will now calculate the EMF obtained from the velocity and magnetic fields composed of the four eigen modes associated with the calculated frequencies (3.8). We start by noting, that the separated problem for vertical components of the fluctuations does not introduce eigen modes capable of creating the mean EMF through interactions with other modes and thus we concentrate on the four eigen modes defined by (3.8) only. We will show, that there are certain types of wave interactions in the studied system, which lead to creation of mean EMF, even in the absence of diffusive effects. Let us rewrite (3.8) in the form

$$\omega_{1,2} = \Omega\frac{k_z}{k} \pm \sqrt{\Omega^2\frac{k_z^2}{k^2} + (\langle\mathbf{B}\rangle\cdot\mathbf{k})^2} \stackrel{\text{def}}{=} \omega_{\pm}, \tag{3.9a}$$

$$\omega_{3,4} = -\Omega\frac{k_z}{k} \mp \sqrt{\Omega^2\frac{k_z^2}{k^2} + (\langle\mathbf{B}\rangle\cdot\mathbf{k})^2} = -\omega_{\pm}, \tag{3.9b}$$

where the plus and minus signs are correlated, which defines the frequencies of the Lehnert waves. We also define

$$\Delta\omega = \omega_1 - \omega_2 = \omega_4 - \omega_3 = 2\sqrt{\Omega^2\frac{k_z^2}{k^2} + (\langle\mathbf{B}\rangle\cdot\mathbf{k})^2}. \tag{3.10}$$

Note, that the oscillation frequency difference (difference between the eigen values) $\Delta\omega$ is itself a nonlinear function of $\langle\mathbf{B}\rangle$ due to action of the Lorentz force. The eigen vectors are

$$\hat{\mathbf{u}}^{(1,2)} = -\frac{\omega_{\pm}}{k^3}\mathbf{V}, \qquad \hat{\mathbf{b}}^{(1,2)} = \frac{\langle\mathbf{B}\rangle\cdot\mathbf{k}}{k^3}\mathbf{V}, \tag{3.11a}$$

$$\hat{\mathbf{u}}^{(3,4)} = \frac{\omega_{\pm}}{k^3}\mathbf{V}^*, \qquad \hat{\mathbf{b}}^{(3,4)} = \frac{\langle\mathbf{B}\rangle\cdot\mathbf{k}}{k^3}\mathbf{V}^*, \tag{3.11b}$$

where

$$\mathbf{V} = \mathbf{k}\times(\mathbf{k}\times\hat{\mathbf{e}}_z) + ik\mathbf{k}\times\hat{\mathbf{e}}_z. \tag{3.12}$$

Note, that modes (1) and (2) have an opposite helicity to that of the modes (3) and (4)

$$\begin{aligned}\langle \mathbf{u}'^{(1,2)} \cdot \nabla \times \mathbf{u}'^{(1,2)} \rangle_{\mathbf{x}} &= -\frac{1}{2}\,\Re\mathrm{e}\,[\mathrm{i}\mathbf{k} \cdot (\hat{\mathbf{u}}^{(1,2)*} \times \hat{\mathbf{u}}^{(1,2)})] \\ &= -\frac{1}{2}\left(\frac{\omega_\pm}{k^3}\right)^2 \Re\mathrm{e}\,[\mathrm{i}\mathbf{k}\cdot(\mathbf{V}^* \times \mathbf{V})] = \frac{1}{2}\,\Re\mathrm{e}\,[\mathrm{i}\mathbf{k}\cdot(\hat{\mathbf{u}}^{(3,4)*} \times \hat{\mathbf{u}}^{(3,4)})] \\ &= -\langle \mathbf{u}'^{(3,4)} \cdot \nabla \times \mathbf{u}'^{(3,4)} \rangle_{\mathbf{x}}.\end{aligned} \tag{3.13}$$

The solution at a given $\mathbf{k}$ consists of a linear combination of all four modes

$$\mathbf{u}' = \sum_{j=1}^{4} \mathcal{F}_j(\mathbf{k})\hat{\mathbf{u}}^{(j)} e^{\mathrm{i}(\mathbf{k}\cdot\mathbf{x} - \omega_j t)} + c.\,c., \qquad \mathbf{b}' = \sum_{j=1}^{4} \mathcal{F}_j(\mathbf{k})\hat{\mathbf{b}}^{(j)} e^{\mathrm{i}(\mathbf{k}\cdot\mathbf{x} - \omega_j t)} + c.\,c, \tag{3.14}$$

where the amplitudes $\mathcal{F}_j$, $j = 1, \ldots, 4$ are constants and $+c.\,c.$ denotes addition of a complex conjugate in order for the resulting fields to be real. Now, we can calculate the EMF from linear combinations of the eigen modes (1) and (2) (one of the only two possibilities leading to non-zero EMF, the other being the combination of modes (3) and (4)),

$$\mathbf{u}' = \mathcal{F}_1\mathbf{u}'^{(1)} + \mathcal{F}_2\mathbf{u}'^{(2)} + c.\,c., \qquad \mathbf{b}' = \mathcal{F}_1\mathbf{b}'^{(1)} + \mathcal{F}_2\mathbf{b}'^{(2)} + c.\,c., \tag{3.15}$$

in the following way

$$\begin{aligned}&\boldsymbol{\mathcal{E}}^{(1;2)}(\mathbf{k}) \\ &= 2\,\Re\mathrm{e}\left[\left(\mathcal{F}_1\hat{\mathbf{u}}^{(1)}e^{-\mathrm{i}\omega_1 t} + \mathcal{F}_2\hat{\mathbf{u}}^{(2)}e^{-\mathrm{i}\omega_2 t}\right) \times \left(\mathcal{F}^*{}_1\hat{\mathbf{b}}^{(1)*}e^{\mathrm{i}\omega_1 t} + \mathcal{F}^*{}_2\hat{\mathbf{b}}^{(2)*}e^{\mathrm{i}\omega_2 t}\right)\right] \\ &= 2\,\Re\mathrm{e}\left[\mathcal{F}_1\mathcal{F}^*{}_2\hat{\mathbf{u}}^{(1)} \times \hat{\mathbf{b}}^{(2)*}e^{-\mathrm{i}\Delta\omega t} + \mathcal{F}_2\mathcal{F}^*{}_1\hat{\mathbf{u}}^{(2)} \times \hat{\mathbf{b}}^{(1)*}e^{\mathrm{i}\Delta\omega t}\right].\end{aligned} \tag{3.16}$$

By the use of (3.11a)

$$\hat{\mathbf{u}}^{(1)} \times \hat{\mathbf{b}}^{(2)*} = 2\mathrm{i}\omega_+\frac{k_h^2}{k^5}(\langle\mathbf{B}\rangle \cdot \mathbf{k})\mathbf{k}, \qquad \hat{\mathbf{u}}^{(2)} \times \hat{\mathbf{b}}^{(1)*} = 2\mathrm{i}\omega_-\frac{k_h^2}{k^5}(\langle\mathbf{B}\rangle \cdot \mathbf{k})\mathbf{k}, \tag{3.17}$$

where $k_h^2 = k_x^2 + k_y^2$, and thus

$$\boldsymbol{\mathcal{E}}^{(1;2)}(\mathbf{k}) = -4\Delta\omega\frac{k_h^2}{k^5}(\langle\mathbf{B}\rangle \cdot \mathbf{k})\mathbf{k}\,\Im\mathrm{m}\,[\mathcal{F}_1\mathcal{F}^*{}_2 e^{-\mathrm{i}\Delta\omega t}]. \tag{3.18}$$

Additionally, from linear combination of modes (3) and (4), i.e. $\bar{\mathbf{u}} = \mathcal{F}_3\mathbf{u}^{(3)} + \mathcal{F}_4\mathbf{u}^{(4)} + c.\,c.$, $\bar{\mathbf{b}} = \mathcal{F}_3\mathbf{b}^{(3)} + \mathcal{F}_4\mathbf{b}^{(4)} + c.\,c.$ one gets

$$\boldsymbol{\mathcal{E}}^{(3;4)}(\mathbf{k}) = 4\Delta\omega\frac{k_h^2}{k^5}(\langle\mathbf{B}\rangle \cdot \mathbf{k})\mathbf{k}\,\Im\mathrm{m}\,[\mathcal{F}_4\mathcal{F}^*{}_3 e^{-\mathrm{i}\Delta\omega t}], \tag{3.19}$$

a similar expression for EMF but with opposite sign. No other types of interactions between eigen modes, that is, between modes (1) and (3), (1) and (4), (2) and (3) or

(2) and (4) lead to non-zero EMF, so that only interactions (1)–(2) and (3)–(4) are generative and the total mean EMF (a sum or integral of contributions for different wave vectors **k**) is

$$\begin{aligned}\mathcal{E} &= \int d^3k[\mathcal{E}^{(1;2)}(\mathbf{k}) + \mathcal{E}^{(3;4)}(\mathbf{k})]\\ &= 4\int d^3k\Delta\omega(\mathbf{k})\frac{k_h^2}{k^5}(\langle\mathbf{B}\rangle\cdot\mathbf{k})\mathbf{k}\,\Im\mathrm{m}\left\{[\mathcal{F}_4(\mathbf{k})\mathcal{F}^*{}_3(\mathbf{k}) - \mathcal{F}_1(\mathbf{k})\mathcal{F}^*{}_2(\mathbf{k})]e^{-i\Delta\omega(\mathbf{k})t}\right\}.\end{aligned} \tag{3.20}$$

The situation is very reminiscent of section 7.7 and section 12 in [2]. For two eigen modes generating the EMF there exist two other eigen modes traveling in the opposite directions which produce a compensating EMF of opposite sign and statistically these two contributions should cancel out (at a given **k** this could be achieved e.g. when $\mathcal{F}_3 = \mathcal{F}_2$ and $\mathcal{F}_4 = \mathcal{F}_1$). However, lack of up-down symmetry, i.e. some mechanism for a preferential excitation of modes traveling in one direction, e.g. modes carrying energy downwards, would lead to a non-zero EMF. The need for lack of up-down symmetry was noted previously in the context of the standard, dissipative α-effect mechanism by many authors, e.g. [2, 12] and many others and assumed arbitrarily, since MHD turbulence is typically generated by spatially localized sources and thus the energy has a preferred direction of propagation. In magnetostrophic turbulence, i.e. dominated by magnetic and Coriolis forces, such symmetry-breaking is provided by rotation in conjunction with gravity or density stratification or shear (cf e.g. [13–16]). In the case at hand, the vertical group velocities of all four eigen modes are

$$c_{gz}^{(1)} = \frac{\partial\omega_1}{\partial k_z} = \Omega\frac{k_h^2}{k^3}\left[1 + \frac{2\Omega}{\Delta\omega}\frac{k_z}{k}\left(1 + \frac{k^2}{k_h^2}\frac{\mathbf{k}\cdot\langle\mathbf{B}\rangle k_z\langle B\rangle_z}{\Omega^2\frac{k_z^2}{k^2}}\right)\right] = -\frac{\partial\omega_3}{\partial k_z} = -c_{gz}^{(3)}, \tag{3.21}$$

$$c_{gz}^{(2)} = \frac{\partial\omega_2}{\partial k_z} = \Omega\frac{k_h^2}{k^3}\left[1 - \frac{2\Omega}{\Delta\omega}\frac{k_z}{k}\left(1 + \frac{k^2}{k_h^2}\frac{\mathbf{k}\cdot\langle\mathbf{B}\rangle k_z\langle B\rangle_z}{\Omega^2\frac{k_z^2}{k^2}}\right)\right] = -\frac{\partial\omega_4}{\partial k_z} = -c_{gz}^{(4)}, \tag{3.22}$$

thus there always exists a pair of modes which carry energy in the direction of rotation vector (say upwards) and another pair which carry the energy in the opposite direction (downwards).

Of course in the real situation, when dissipative effects are present, the eigen waves must decay. The decay can be very slow if diffusion is weak, such as e.g. in the case of interstellar plasma, when the characteristic magnetic diffusion time is much longer than the age of the Universe. In such a case the mean EMF decays due to viscosity, but once the mean magnetic field is generated at $\eta \approx 0$ by the non-stationary mechanism of interactions between distinct waves, even when the EMF is finally suppressed by viscosity, the magnetic field may survive as a fossil field. This is because there can exist force-free modes, for which the mean currents $\langle \mathbf{J} \rangle$ flow along the mean magnetic field lines, thus the Lorentz force $\langle \mathbf{J} \rangle \times \langle \mathbf{B} \rangle = 0$ vanishes and there is no effective mechanism of their energy decay. Such a situation was considered in stationary turbulence by Mizerski and Moffatt [17] (see also [2]) with yet another mechanism of mean EMF generation through viscous phase shift between velocity and magnetic fields of a single Lehnert wave.

Nevertheless, in the absence of any forcing of the diffusion, $\nu \neq 0$ and $\eta \neq 0$, this leads to weakening and eventually vanishing of the mean EMF. However, if the flow is forced (stirred) resonance effects supply the selected waves with energy, which overcome the diffusion. For example, we can introduce non-zero but small magnetic diffusion η, and assume it is weak in the sense that $E_\eta = \eta/\Omega L^2 \ll 1$ is small. In such a case the eigenvalues are modified

$$\mathring{\omega}_{1,2} = \omega_\pm \mp \mathrm{i}\eta k^2 \frac{\omega_\mp}{\Delta\omega}, \qquad \mathring{\omega}_{3,4} = -\omega_\pm \mp \mathrm{i}\eta k^2 \frac{\omega_\mp}{\Delta\omega}, \tag{3.23}$$

(recall, that $\omega_{1,2} = \omega_\pm$ and $\omega_{3,4} = -\omega_\pm$) and the two eigen vectors associated with the eigen values $\mathring{\omega}_1$ and $\mathring{\omega}_2$ are given by

$$\hat{\mathbf{u}}^{(1,2)} = \frac{\omega_\pm}{\mathcal{B}}\left(1 \mp \mathrm{i}\frac{\eta k^2}{\Delta\omega}\right)\hat{\mathbf{b}}^{(1,2)}, \qquad \hat{\mathbf{b}}^{(1,2)} = \frac{\mathcal{B}}{k^3}\mathbf{V} + \mathcal{O}(E_\eta). \tag{3.24}$$

When the flow is forced at a fixed frequency, say $\bar{\omega}_1 > 0$, only those wave vectors in the system are picked up, which correspond to the frequency of oscillation $\bar{\omega}_1$ through the dispersion relation (3.9), i.e. the wave vectors of undamped modes belong to the set

$$\mathscr{K} = \left\{ \mathbf{k}\colon\ \Omega\frac{k_z}{k} + \sqrt{\Omega^2\frac{k_z^2}{k^2} + (\langle\mathbf{B}\rangle\cdot\mathbf{k})^2} \approx \bar{\omega}_1, \right.$$
$$\left. \text{or} \ -\Omega\frac{k_z}{k} + \sqrt{\Omega^2\frac{k_z^2}{k^2} + (\langle\mathbf{B}\rangle\cdot\mathbf{k})^2} \approx \bar{\omega}_1 \right\}. \tag{3.25}$$

We turn to such a situation in the next section.

3.2 Waves forced at two distinct frequencies

We now consider the case of stirred MHD, $\mathbf{f} \neq 0$, and suppose that the stirring force $\mathbf{f}$ is reflectionally symmetric, thus does not introduce helicity into the flow, $\langle \mathbf{f} \cdot \nabla \times \mathbf{f} \rangle = \mathbf{0}$. Furthermore, we will consider the effect of two Fourier modes of the forcing $\mathbf{f}$, with the same wave vector but distinct frequencies of time oscillation,

$$\mathbf{f} = \mathfrak{Re}\,\{\hat{\mathbf{f}}_1 \exp[\mathrm{i}(\mathbf{k}\cdot\mathbf{x} - \bar{\omega}_1 t)]\} + \mathfrak{Re}\,\{\hat{\mathbf{f}}_2 \exp[\mathrm{i}(\mathbf{k}\cdot\mathbf{x} - \bar{\omega}_2 t)]\}, \tag{3.26}$$

easily generalizable to the case of many temporal Fourier modes. Therefore, the linearized equations (3.4a–c) admit solutions in the form of forced waves

$$(\mathbf{u},\,\mathbf{b},\,\Pi') = \mathfrak{Re}\Big\{(\hat{\mathbf{u}}_1,\,\hat{\mathbf{b}}_1,\,\hat{\Pi}_1)\exp\left[\mathrm{i}\mathbf{k}\cdot(\mathbf{x}-\langle\mathbf{U}\rangle t)-\mathrm{i}\varpi_1 t\right]\Big\} + \mathfrak{Re}\Big\{(\hat{\mathbf{u}}_2,\,\hat{\mathbf{b}}_2,\,\hat{\Pi}_2)\exp\left[\mathrm{i}\mathbf{k}\cdot(\mathbf{x}-\langle\mathbf{U}\rangle t)-\mathrm{i}\varpi_2 t\right]\Big\}, \tag{3.27}$$

which oscillate at the frequencies of the forcing in a frame moving with the mean velocity $\langle\mathbf{U}\rangle$ and do not decay even when $\eta \neq 0$. The divergence conditions (3.4c) imply that

$$\mathbf{k}\cdot\hat{\mathbf{u}} = \mathbf{k}\cdot\hat{\mathbf{b}} = 0, \tag{3.28}$$

so the waves are transverse.

The compact treatment of [2] (chapter 12.2 of that book) is followed here for ease of comparison. Substitution first in (3.4b) gives the familiar relationship between $\hat{\mathbf{b}}$ and $\hat{\mathbf{u}}$, *viz.*

$$\hat{\mathbf{b}}_j = -(\varpi_j + \mathrm{i}\eta k^2)^{-1}(\mathbf{k}\cdot\langle\mathbf{B}\rangle)\hat{\mathbf{u}}_j, \tag{3.29}$$

and then substitution in (3.4a) and rearrangement of the terms gives

$$-\mathrm{i}\sigma_j\hat{\mathbf{u}}_j + 2\boldsymbol{\Omega}\times\hat{\mathbf{u}}_j = -\mathrm{i}\,\mathbf{k}\hat{\Pi}_j + \hat{\mathbf{f}}_j, \tag{3.30a}$$

where

$$\sigma_j = \varpi_j + \mathrm{i}\nu k^2 - (\varpi_j + \mathrm{i}\eta k^2)^{-1}(\mathbf{k}\cdot\langle\mathbf{B}\rangle)^2, \tag{3.30b}$$

and $j = 1, 2$. Let us consider high-conductivity low-viscosity limit,

$$\eta k^2 \ll |\sigma_j|, \qquad \nu k^2 \ll |\sigma_j|, \tag{3.31}$$

in which case the expression (3.2) can be simplified to yield

$$\mathfrak{Re}\,\sigma_j \approx \varpi_j - \frac{(\mathbf{k}\cdot\langle\mathbf{B}\rangle)^2}{\varpi_j}, \qquad \mathfrak{Im}\,\sigma_j \approx \nu k^2 + \eta k^2\frac{(\mathbf{k}\cdot\langle\mathbf{B}\rangle)^2}{\varpi_j^2}. \tag{3.32}$$

To get the velocity field, we take the double cross-product of (3.30a) with $\mathbf{k}$, that is, we apply to both sides $\mathbf{k}\times\mathbf{k}\times$; since $\mathbf{k}\cdot\hat{\mathbf{u}} = 0$, this allows to get

$$\hat{\mathbf{u}}_j = D_j^{-1}\left[2(\mathbf{k}\cdot\boldsymbol{\Omega})\mathbf{k}\times\hat{\mathbf{f}}_j + \mathrm{i}\sigma_j k^2\hat{\mathbf{f}}_j\right], \tag{3.33}$$

where

$$D_j = \sigma_j^2 k^2 - 4(\mathbf{k}\cdot\boldsymbol{\Omega})^2. \tag{3.34}$$

A resonant response is obtained when the frequency of the forcing ϖ_j coincides with one of the natural (eigen) frequencies of the undamped system (for the given wave vector $\mathbf{k}$), that is, the frequencies of the Lehnert waves given in (3.9). This happens when

$$\mathfrak{Re}\sigma_j \approx \pm 2(\mathbf{k}\cdot\boldsymbol{\Omega})/k, \tag{3.35}$$

which corresponds to minimum value of $|D_j|$. In the latter equation the + sign corresponds to the eigen frequencies ω_1 and ω_2 in (3.9), whereas the − sign corresponds to the other two eigen frequencies, ω_3 and ω_4.

We now calculate the helicity of the forced velocity field $\mathbf{u}$

$$\begin{aligned}\langle \mathbf{u}\cdot\nabla\times\mathbf{u}\rangle &= -\frac{1}{2}\,\Re\mathfrak{e}\,[\mathrm{i}\mathbf{k}\cdot(\hat{\mathbf{u}}_1^*\mathrm{e}^{\mathrm{i}\bar{\omega}_1 t}+\hat{\mathbf{u}}_2^*\mathrm{e}^{\mathrm{i}\bar{\omega}_2 t})\times(\hat{\mathbf{u}}_1\mathrm{e}^{-\mathrm{i}\bar{\omega}_1 t}+\hat{\mathbf{u}}_2\mathrm{e}^{-\mathrm{i}\bar{\omega}_2 t})]\\ &= -\,\Re\mathfrak{e}\,[\mathrm{i}\mathbf{k}\cdot(\hat{\mathbf{u}}_1^*\times\hat{\mathbf{u}}_2\mathrm{e}^{\mathrm{i}\Delta\bar{\omega}t})]-\frac{1}{2}\,\Re\mathfrak{e}\,[\mathrm{i}\mathbf{k}\cdot(\hat{\mathbf{u}}_1^*\times\hat{\mathbf{u}}_1)+\mathrm{i}\mathbf{k}\cdot(\hat{\mathbf{u}}_2^*\times\hat{\mathbf{u}}_2)],\end{aligned} \tag{3.36}$$

where $\Delta\bar{\omega} = \bar{\omega}_1 - \bar{\omega}_2$ is the difference between the two frequencies of the forcing. Since $\hat{\mathbf{u}}_j$ and $\hat{\mathbf{u}}_j^*$ for $j = 1, 2$ are perpendicular to $\mathbf{k}$, it follows that $(\hat{\mathbf{u}}_i^* \times \hat{\mathbf{u}}_j) \parallel \mathbf{k}$, so that it is enough to calculate $\hat{\mathbf{u}}_i^* \times \hat{\mathbf{u}}_j$. The last two terms in the expression (3.36) correspond to the well-known mechanism of mean EMF generation from waves forced by just one frequency, either $\bar{\omega}_1$ or $\bar{\omega}_2$, for which the solution, and in particular the helicity, was provided in formula (12.96) of chapter 12.6 in [2], and it is recalled here for completeness

$$\mathrm{i}\hat{\mathbf{u}}_j^*\times\hat{\mathbf{u}}_j = 4\,|\,D_j\,|^{-2}\,(\mathbf{k}\cdot\boldsymbol{\Omega})k^2\,\Re\mathfrak{e}\,\sigma_j\,\big|\,\hat{\mathbf{f}}_j\,\big|^2\,\mathbf{k}, \tag{3.37}$$

for $j = 1, 2$. The first term in (3.36) describes the non-equilibrium mechanism based on statistical non-stationarity of the stirring force, i.e. interactions between waves with distinct frequencies and can be calculated to yield

$$\mathrm{i}\hat{\mathbf{u}}_1^*\times\hat{\mathbf{u}}_2 = 2(D_1^*D_2)^{-1}(\mathbf{k}\cdot\boldsymbol{\Omega})k^2(\sigma_1^*+\sigma_2)\left(\hat{\mathbf{f}}_1^*\cdot\hat{\mathbf{f}}_2\right)\mathbf{k}. \tag{3.38}$$

In the two latter formulae we have utilized the assumption of reflectional symmetry of $\mathbf{f}$, which implies

$$\left(\hat{\mathbf{f}}_1\mathrm{e}^{\mathrm{i}\bar{\omega}_1 t}+\hat{\mathbf{f}}_2\mathrm{e}^{\mathrm{i}\bar{\omega}_2 t}\right)\cdot\left[\mathbf{k}\times\left(\hat{\mathbf{f}}_1^*\mathrm{e}^{-\mathrm{i}\bar{\omega}_1 t}+\hat{\mathbf{f}}_2^*\mathrm{e}^{-\mathrm{i}\bar{\omega}_2 t}\right)\right]=0, \tag{3.39}$$

and hence

$$\hat{\mathbf{f}}_i^*\times\mathbf{f}_j = 0, \tag{3.40}$$

for all combinations of $i = 1, 2$ and $j = 1, 2$. Furthermore, in the weak resistivity limit we get from (3.32)

$$\sigma_1^*+\sigma_2 \approx (\bar{\omega}_1+\bar{\omega}_2)\left[1-\frac{(\mathbf{k}\cdot\langle\mathbf{B}\rangle)^2}{\bar{\omega}_1\bar{\omega}_2}+\mathrm{i}\eta k^2\frac{\Delta\bar{\omega}(\mathbf{k}\cdot\langle\mathbf{B}\rangle)^2}{\bar{\omega}_1^2\bar{\omega}_2^2}\right]. \tag{3.41}$$

In order to understand the physical picture it is helpful to consider the limit of weak seed magnetic field $|\,\mathbf{k}\cdot\langle\mathbf{B}\rangle\,| \ll \bar{\omega}_1, \bar{\omega}_2$, when the helicity is expressed by terms proportional to $(\bar{\omega}_i + \bar{\omega}_j)(\mathbf{k}\cdot\boldsymbol{\Omega})$. This means, that the helical forced waves propagate either 'upwards' or 'downwards' relative to the direction of $\boldsymbol{\Omega}$. In order to have a non-zero net helicity in a superposition of such waves, the forcing must be such as to preferentially excite waves of either positive or negative helicity, so that the net helicity flux is non-zero. We assume that such symmetry breaking exists in the present context, and this allows us to focus on waves of, say, positive helicity, corresponding to the choice $\sigma_j = -2(\mathbf{k}\cdot\mathbf{\Omega})/k$ in (3.35); however, we stress, that the weak seed field limit is *not* considered in the following, until section 3.2.2 and the following formulae are valid for arbitrary strength of the mean field.

We can now write down the general form of the mean EMF generated by waves forced at two distinct frequencies

$$\begin{aligned}\hat{\boldsymbol{\mathcal{E}}}(\mathbf{k}) = \langle \mathbf{u}\times\mathbf{b}\rangle = \frac{1}{2}\mathfrak{Re}\left[(\hat{\mathbf{u}}_1 e^{-i\bar{\omega}_1 t} + \hat{\mathbf{u}}_2 e^{-i\bar{\omega}_2 t})\times\left(\hat{\mathbf{b}}_1^* e^{i\bar{\omega}_1 t} + \hat{\mathbf{b}}_2^* e^{i\bar{\omega}_2 t}\right)\right] \\ = \hat{\boldsymbol{\mathcal{E}}}_{(1;2)}(\mathbf{k}) + \hat{\boldsymbol{\mathcal{E}}}_1(\mathbf{k}) + \hat{\boldsymbol{\mathcal{E}}}_2(\mathbf{k}),\end{aligned} \tag{3.42}$$

where

$$\hat{\boldsymbol{\mathcal{E}}}_j(\mathbf{k}) = \frac{1}{2}\mathfrak{Re}\left(\hat{\mathbf{u}}_j\times\hat{\mathbf{b}}_j^*\right) \tag{3.43}$$

for $j = 1, 2$ is the standard single-frequency contribution based on self-interactions between the kinetic and magnetic components of a single wave and the non-equilibrium effect is contained in

$$\hat{\boldsymbol{\mathcal{E}}}_{(1;2)}(\mathbf{k}) = \frac{1}{2}\mathfrak{Re}\left(\hat{\mathbf{u}}_2\times\hat{\mathbf{b}}_1^* e^{i\Delta\bar{\omega}t} + \hat{\mathbf{u}}_1\times\hat{\mathbf{b}}_2^* e^{-i\Delta\bar{\omega}t}\right). \tag{3.44}$$

By the use of (3.29) and in the weak resistivity limit $\eta k^2 \ll |\sigma_j|$ we can simplify the latter expression to

$$\begin{aligned}\hat{\boldsymbol{\mathcal{E}}}_{(1;2)}(\mathbf{k}) = -\frac{\mathbf{k}\cdot\langle\mathbf{B}\rangle}{2\bar{\omega}_1^2\bar{\omega}_2^2}\Big[\bar{\omega}_1\bar{\omega}_2\Delta\bar{\omega}\,\mathfrak{Im}\,(i\hat{\mathbf{u}}_1^*\times\hat{\mathbf{u}}_2 e^{i\Delta\bar{\omega}t}) \\ -\eta k^2\left(\bar{\omega}_1^2 + \bar{\omega}_2^2\right)\mathfrak{Re}\left(i\hat{\mathbf{u}}_1^*\times\hat{\mathbf{u}}_2 e^{i\Delta\bar{\omega}t}\right)\Big] + \mathcal{O}\left(\frac{\eta^2k^4}{\bar{\omega}_1^2}, \frac{\eta^2k^4}{\bar{\omega}_2^2}\right).\end{aligned} \tag{3.45}$$

Once again, for completeness we rewrite the single-wave EMF from equation (12.84) in chapter 12.6 of [2],

$$\hat{\boldsymbol{\mathcal{E}}}_j(\mathbf{k}) = \eta k^2\frac{\mathbf{k}\cdot\langle\mathbf{B}\rangle}{2\bar{\omega}_j^2}\left(i\hat{\mathbf{u}}_j^*\times\hat{\mathbf{u}}_j\right) + \mathcal{O}\left(\frac{\eta^3k^6}{\bar{\omega}_j^3}\right). \tag{3.46}$$

where only the leading order term in η was retained. Note that the second term in (3.45) is of the same order in terms of η and of similar form to the single-frequency EMF in (3.46).

The next step is to assume that the frequencies of the forcing are close enough, that $\Delta\bar{\omega} \ll \bar{\omega}_1, \bar{\omega}_2$ and the slow oscillations can be effectively removed by approximating the exponentials in (3.45) by $\exp(i\Delta\bar{\omega}t) \approx 1 + i\Delta\bar{\omega}t$ and then to consider the conditions of resonance defined in (3.35). This will be done in the following sections.

3.2.1 Mean EMF induced by forced beating waves

As argued in the introduction in highly conductive plasma, when the characteristic magnetic diffusion times are very large, it seems legitimate to consider only the leading order terms within the low-resistivity limit (3.31). In such a case the mean EMF greatly simplifies. Using (3.45) and (3.41), this is given by

$$\begin{aligned}\hat{\mathcal{E}}_{(1;2)}(\mathbf{k}) &= -\frac{\Delta\varpi(\mathbf{k}\cdot\boldsymbol{\Omega})\mathbf{k}\cdot\langle\mathbf{B}\rangle}{\varpi_1\varpi_2}k^3\,\Im\mathrm{m}\left[\frac{\sigma_1^*+\sigma_2}{D_1^*D_2}\hat{\mathbf{f}}_1^*\cdot\hat{\mathbf{f}}_2e^{\mathrm{i}\Delta\varpi t}\right]\frac{\mathbf{k}}{k}\\ &= -\Delta\varpi(\varpi_1+\varpi_2)\left(\frac{\mathbf{k}\cdot\boldsymbol{\Omega}}{k}\right)\mathbf{k}\cdot\langle\mathbf{B}\rangle[\varpi_1\varpi_2-(\mathbf{k}\cdot\langle\mathbf{B}\rangle)^2]\,\Im\mathrm{m}\left[\frac{\hat{\mathbf{f}}_1^*\cdot\hat{\mathbf{f}}_2e^{\mathrm{i}\Delta\varpi t}}{\varpi_1^2D_1^*\varpi_2^2D_2/k^4}\right]\frac{\mathbf{k}}{k},\end{aligned}\tag{3.47}$$

and from (3.32) and (3.34)

$$\begin{aligned}\frac{\varpi_j^2D_j}{k^2} &= \varpi_j^2\sigma_j^2-4\varpi_j^2\left(\frac{\mathbf{k}\cdot\boldsymbol{\Omega}}{k}\right)^2\\ &\approx\left[\varpi_j^2-(\mathbf{k}\cdot\langle\mathbf{B}\rangle)^2\right]^2-4\varpi_j^2\left(\frac{\mathbf{k}\cdot\boldsymbol{\Omega}}{k}\right)^2-\left[\frac{\nu k^2}{\varpi_j}\varpi_j^2+\frac{\eta k^2}{\varpi_j}(\mathbf{k}\cdot\langle\mathbf{B}\rangle)^2\right]^2\\ &\quad+2\mathrm{i}\left[\varpi_j^2-(\mathbf{k}\cdot\langle\mathbf{B}\rangle)^2\right]\left[\frac{\nu k^2}{\varpi_j}\varpi_j^2+\frac{\eta k^2}{\varpi_j}(\mathbf{k}\cdot\langle\mathbf{B}\rangle)^2\right].\end{aligned}\tag{3.48}$$

We have retained the contribution from small viscosity ν and resistivity η in the latter expression, because it appears in the denominator of (3.47), and as we will see below it is necessary to regularize the expression for the EMF when the forcing frequency is in resonance with one of the eigen frequencies of the system (frequency of the Lehnert wave). The resonance intensifies the EMF, which in absence of diffusion blows up to infinity.

We will now consider an asymptotic limit of very small difference between the frequencies of the forcing ϖ_1 and ϖ_2

$$\Delta\varpi=\varpi_1-\varpi_2\ll\varpi_1,\varpi_2,\tag{3.49}$$

so that the time oscillations of the EMF are slow, which allows us to simplify the expression for the mean EMF at leading order even more

$$\hat{\mathcal{E}}_{(1;2)}(\mathbf{k})\approx-\frac{2\Delta\varpi\left(\frac{\mathbf{k}\cdot\boldsymbol{\Omega}}{k}\right)\varpi_1[\varpi_1^2-(\mathbf{k}\cdot\langle\mathbf{B}\rangle)^2]\mathbf{k}\cdot\langle\mathbf{B}\rangle}{\varpi_1^4\mid D_1\mid^2/k^4}\,\Im\mathrm{m}\left(\hat{\mathbf{f}}_1^*\cdot\hat{\mathbf{f}}_2\right)\frac{\mathbf{k}}{k},\tag{3.50a}$$

$$\begin{aligned}\frac{\varpi_1^4|D_1|^2}{k^4} &\approx\left\{\left[\varpi_1^2-(\mathbf{k}\cdot\langle\mathbf{B}\rangle)^2\right]^2-4\varpi_1^2\left(\frac{\mathbf{k}\cdot\boldsymbol{\Omega}}{k}\right)^2-\left[\frac{\nu k^2}{\varpi_1}\varpi_1^2+\frac{\eta k^2}{\varpi_1}(\mathbf{k}\cdot\langle\mathbf{B}\rangle)^2\right]^2\right\}^2\\ &\quad+4\left[\varpi_1^2-(\mathbf{k}\cdot\langle\mathbf{B}\rangle)^2\right]^2\left[\frac{\nu k^2}{\varpi_1}\varpi_1^2+\frac{\eta k^2}{\varpi_1}(\mathbf{k}\cdot\langle\mathbf{B}\rangle)^2\right]^2.\end{aligned}\tag{3.50b}$$

Such a small difference between the frequencies of waves of the same type corresponds to the effect of 'beat' in acoustics, which is why we adopt the term '*beating waves*' in the considered setting. On the other hand, in the weak diffusion limit (3.31) the single-mode EMF (3.46) generated by self-interactions within one forced wave takes the leading order form

$$\hat{\mathcal{E}}_j(\mathbf{k}) = \frac{2\eta k^2\left(\dfrac{\mathbf{k}\cdot\boldsymbol{\Omega}}{k}\right)\bar{\omega}_j\left[\bar{\omega}_j^2 - (\mathbf{k}\cdot\langle\mathbf{B}\rangle)^2\right]\mathbf{k}\cdot\langle\mathbf{B}\rangle}{\bar{\omega}_j^4\,|D_j|^2/k^4}\,|\hat{\mathbf{f}}_j|^2\,\frac{\mathbf{k}}{k}. \tag{3.51}$$

The results allow us to compare the magnitudes of both contributions to the total EMF, i.e. the non-equilibrium contribution $\hat{\mathcal{E}}_{(1;2)}(\mathbf{k})$ with the standard one $\hat{\mathcal{E}}_j(\mathbf{k})$,

$$\frac{\left\|\hat{\mathcal{E}}_j(\mathbf{k})\right\|}{\left\|\hat{\mathcal{E}}_{(1;2)}(\mathbf{k})\right\|} = \frac{\eta k^2\,|\hat{\mathbf{f}}_j|^2}{\Delta\bar{\omega}\,\Im\mathrm{m}\left(\hat{\mathbf{f}}_1^*\cdot\hat{\mathbf{f}}_2\right)} = \frac{\eta k^2}{\Delta\bar{\omega}}\frac{|\hat{\mathbf{f}}_j|^2}{\left|\Im\mathrm{m}\left(\hat{\mathbf{f}}_1^*\cdot\hat{\mathbf{f}}_2\right)\right|}. \tag{3.52}$$

The EMF from beating waves in (3.50) is therefore dominant over the EMF induced by a single wave (3.51) if

$$\eta k^2 \ll \Delta\bar{\omega} \ll \bar{\omega}_1,\ \bar{\omega}_2, \tag{3.53}$$

is satisfied. In other words, in order for the beating effect to be dominant, the frequency difference $\Delta\bar{\omega}$ must be small, so that the time oscillations of the EMF are slow, but not too small, that is, still significantly greater than diffusion, ηk^2.

The EMF $\hat{\mathcal{E}}_{(1;2)}(\mathbf{k})$ (3.50) is present even in the perfectly conducting limit $\eta = 0$. In this limit, it is the interactions between beating waves forced at different but close frequencies, say $\mathbf{u}_1$ and $\mathbf{b}_2$, that produce a non-zero mean $\langle\mathbf{u}_1\times\mathbf{b}_2\rangle$. This is an interesting situation, because the $\eta\to 0$ limit usually possesses severe difficulties in dynamo theory. Here however, we have a well-defined α-effect. We shall explore the consequences in section 3.2.2. It is of interest to comment on the possible case of resonant response of the system defined by (3.35). In an ideal fluid a perfect resonance coincidence of the forcing and eigen frequencies is not possible, as it would lead to an infinite amplitude of the forced wave. This is why the regularizing effect of even very weak resistivity is important, since in real fluids the resonance excites (feeds with energy) modes with maximal amplitudes in the system. Denoting the eigen frequency of the system, i.e. the frequency of Lehnert waves by ω_L (i.e. ω_L is equal to any of the four linear eigen values given in (3.9)), we can consider modes satisfying

$$\bar{\omega}_1 = \omega_L, \qquad \bar{\omega}_2 = \omega_L - \Delta\bar{\omega}, \tag{3.54a}$$

$$\Delta\bar{\omega} = \bar{\omega}_1 - \bar{\omega}_2 \ll \bar{\omega}_1,\ \bar{\omega}_2. \tag{3.54b}$$

Since

$$\omega_L^2 - (\mathbf{k}\cdot\langle\mathbf{B}\rangle)^2 = \pm 2\frac{\mathbf{k}\cdot\boldsymbol{\Omega}}{k}\omega_L, \tag{3.55}$$

(where the + sign corresponds to the ω_1, ω_2 eigen frequencies and the − sign to ω_3, ω_4, cf (3.9)) the mean EMF under such resonance conditions, at the leading order takes the form

$$\hat{\mathcal{E}}_{(1;2)}(\mathbf{k}) \approx \mp\frac{\Delta\bar{\omega}}{\eta k^2}\frac{4\eta k^2\left(\frac{\mathbf{k}\cdot\boldsymbol{\Omega}}{k}\right)^2\bar{\omega}_L^2\mathbf{k}\cdot\langle\mathbf{B}\rangle}{\bar{\omega}_L^4\mid D_1\mid^2/k^4}\,\Im\mathrm{m}\left(\hat{\mathbf{f}}_1^*\cdot\hat{\mathbf{f}}_2\right)\frac{\mathbf{k}}{k}, \tag{3.56}$$

$$\hat{\mathcal{E}}_j(\mathbf{k}) = \pm\frac{4\eta k^2\left(\frac{\mathbf{k}\cdot\boldsymbol{\Omega}}{k}\right)^2\omega_L^2\mathbf{k}\cdot\langle\mathbf{B}\rangle}{\bar{\omega}_L^4\mid D_1\mid^2/k^4}\left|\,\hat{\mathbf{f}}_j\right|^2\frac{\mathbf{k}}{k}, \tag{3.57}$$

where the denominators are now

$$\begin{aligned}\frac{\bar{\omega}_L^4\mid D_1\mid^2}{k^4} &\approx \left[\frac{\nu k^2}{\bar{\omega}_L}\bar{\omega}_L^2 + \frac{\eta k^2}{\bar{\omega}_L}(\mathbf{k}\cdot\langle\mathbf{B}\rangle)^2\right]^2\left\{\left[\frac{\nu k^2}{\bar{\omega}_L}\bar{\omega}_L^2 + \frac{\eta k^2}{\bar{\omega}_L}(\mathbf{k}\cdot\langle\mathbf{B}\rangle)^2\right]^2 + 16\left(\frac{\mathbf{k}\cdot\boldsymbol{\Omega}}{k}\right)^2\omega_L^2\right\}\\ &\approx 16\left(\frac{\mathbf{k}\cdot\boldsymbol{\Omega}}{k}\right)^2\omega_L^2\left[\frac{\nu k^2}{\bar{\omega}_L}\bar{\omega}_L^2 + \frac{\eta k^2}{\bar{\omega}_L}(\mathbf{k}\cdot\langle\mathbf{B}\rangle)^2\right]^2.\end{aligned} \tag{3.58}$$

Once again, the upper sign in (3.56) corresponds to $\omega_L = \omega_1$ or $\omega_L = \omega_2$, whereas the lower sign to $\omega_L = \omega_3$ or $\omega_L = \omega_4$. Therefore, for resonances, the EMF is greatly amplified, since the denominator becomes small, $\mid D_1\mid^2 = \mathcal{O}(\nu^2k^4/\bar{\omega}_L^2, \eta^2k^4/\bar{\omega}_L^2, \nu\eta k^4/\bar{\omega}_L^2)$. However, it is of interest to point out, that in a real fluid the resonant response may be localized in thin layers of enhanced diffusion (such as current-sheets) occurring for short-wavelength waves; then the global dynamical significance of the resonances as in (3.54) seems to be reduced.

3.2.2 Dynamo action from beating waves

We will now study in detail the possibility of dynamo action from initially weak seed field, with the EMF given in (3.50), that is, we concentrate here solely on the effect of the EMF from interactions of beating waves. First let us note that a spectrum of waves of different **k** will provide additive contributions to the total EMF $\mathcal{E}$ which then in the limit $\Delta\bar{\omega} \gg \eta k^2$ takes the form

$$\mathcal{E} \approx -2\frac{\Delta\bar{\omega}}{\bar{\omega}_1}\int\frac{\left(\frac{\mathbf{k}\cdot\boldsymbol{\Omega}}{k}\right)\frac{\mathbf{k}}{k}\mathbf{k}\cdot\langle\mathbf{B}\rangle}{\left[\bar{\omega}_1^2 - 4\left(\frac{\mathbf{k}\cdot\boldsymbol{\Omega}}{k}\right)^2\right]^2 + 2\nu^2k^4\left[\bar{\omega}_1^2 + 4\left(\frac{\mathbf{k}\cdot\boldsymbol{\Omega}}{k}\right)^2\right]}\,\Im\mathrm{m}\left(\hat{\mathbf{f}}_1^*\cdot\hat{\mathbf{f}}_2\right)d^3k. \tag{3.59}$$

It is evident, that without any phase shift between the $\mathbf{f}_1$ and $\mathbf{f}_2$ modes the imaginary part of $\hat{\mathbf{f}}_1^*\cdot\hat{\mathbf{f}}_2$ must vanish and hence the EMF is not generated by interactions of beating forced waves. Therefore, in order for the non-stationary α-effect to efficiently operate it is crucial that the two oscillatory components of the forcing, which oscillate at two different but close frequencies are also out of phase from the initial instant. To fix ideas we can assume that the two forcing components $\hat{\mathbf{f}}_1$ and $\hat{\mathbf{f}}_2$ have the same amplitudes and a phase shift of $\pi/2$ i.e. $\hat{\mathbf{f}}_1 = \mathrm{i}\hat{\mathcal{F}}(\mathbf{k})$, $\hat{\mathbf{f}}_2 = \hat{\mathcal{F}}(\mathbf{k})$ and $\hat{\mathcal{F}}(\mathbf{k}) \in \mathbb{R}$, which in particular implies that the diffusive EMF (3.56) and the non-equilibrium one (3.51) have the same signs and differ only by the factor of $\Delta\varpi/\eta k^2$. Consequently,

$$\mathcal{E}_i = \alpha_{ij}\langle B\rangle_j, \tag{3.60}$$

where

$$\alpha_{ij} \approx 2\frac{\Delta\varpi}{\varpi_1}\Omega \int \frac{k\hat{\mathcal{F}}(\mathbf{k})^2}{\left(\varpi_1^2 - 4\Omega^2X^2\right)^2 + 2\nu^2k^4\left(\varpi_1^2 + 4\Omega^2X^2\right)} \frac{k_ik_jk_z}{k^3}\mathrm{d}^3k, \tag{3.61}$$

with

$$X = \frac{k_z}{k}. \tag{3.62}$$

This is an α-effect, in general anisotropic.

Next, as already explained below (3.20) and (3.41), we must break somehow the up-down symmetry of the Fourier spectrum, in order for the α-effect to operate. We consider here the clearest and simplest case, which also gives a most evident α-effect, of turbulence fully dominated by upward propagating waves, i.e. the spectrum of the downward propagating waves vanishes and to fix ideas we also assume $\bar{\varpi}_1 > 0$. It follows, that any spectral d^3k integral takes the form

$$\int F(\mathbf{k})\mathrm{d}^3k = \int k^2\mathrm{d}k \int \mathrm{d}\mathring{\Omega}_+ F(\mathbf{k}) = \int k^2\mathrm{d}k \int_0^{2\pi} \mathrm{d}\varphi \int_0^{\pi/2} \sin\theta\mathrm{d}\theta F(k, \cos\theta, \varphi), \tag{3.63}$$

where $\mathring{\Omega}_+$ denotes the upper half of a solid angle for $0 \leqslant k_z = k\cos\theta \leqslant k$, θ is the meridional angle in spherical coordinates (k, θ, φ) with $\theta = 0$ on the axis of rotation $\mathbf{\Omega} = \Omega\hat{\mathbf{e}}_z$; it follows, that $X = \cos\theta$. For simplicity, let us suppose that the force spectrum $\hat{\mathcal{F}}(\mathbf{k})^2 = \hat{\mathcal{F}}^2(k, \theta)$ is symmetric about the direction of $\mathbf{\Omega}$; then α_{ij} is also axisymmetric. Making use of

$$\begin{aligned}\int \frac{k_ik_jk_k}{k^3} f(\cos\theta)\mathrm{d}\mathring{\Omega}_+ = \pi\left(\delta_{i3}\delta_{jk} + \delta_{j3}\delta_{ik} + \delta_{k3}\delta_{ij}\right)\int_0^1 f(X)X(1 - X^2)\mathrm{d}X \\ + \pi\delta_{i3}\delta_{j3}\delta_{k3}\int_0^1 f(X)X(5X^2 - 3)\mathrm{d}X,\end{aligned} \tag{3.64}$$

which can be derived in a straightforward manner, we arrive at

$$\begin{aligned}
\alpha_{ij} &\approx 2\frac{\Delta\bar{\omega}}{\bar{\omega}_1}\Omega\int \mathrm{d}k\int_0^{2\pi}\mathrm{d}\varphi\int_0^1 \mathrm{d}X\frac{\hat{\mathcal{F}}^2(k,\theta)k_ik_jk_z}{(\bar{\omega}_1^2-4\Omega^2X^2)^2+2\nu^2k^4(\bar{\omega}_1^2+4\Omega^2X^2)}\\
&\approx \delta_{ij}2\pi\frac{\Delta\bar{\omega}}{\bar{\omega}_1}\Omega\int \mathrm{d}k\int_0^1 \mathrm{d}X\frac{k^3\hat{\mathcal{F}}^2(k,\theta)X(1-X^2)}{(\bar{\omega}_1^2-4\Omega^2X^2)^2+2\nu^2k^4(\bar{\omega}_1^2+4\Omega^2X^2)}\\
&+\delta_{i3}\delta_{j3}2\pi\frac{\Delta\bar{\omega}}{\bar{\omega}_1}\Omega\int \mathrm{d}k\int_0^1 \mathrm{d}X\frac{k^3\hat{\mathcal{F}}^2(k,\theta)X(3X^2-1)}{(\bar{\omega}_1^2-4\Omega^2X^2)^2+2\nu^2k^4(\bar{\omega}_1^2+4\Omega^2X^2)}.
\end{aligned}\tag{3.65}$$

In order to obtain a simple estimate of the non-stationary α-effect in the next step we propose the following simplified form of the force spectrum function

$$\hat{\mathcal{F}}^2(k,\theta)=\mathscr{F}^2k^{-n},\tag{3.66}$$

where the units of the constant $\mathscr{F}^2$ are m^{5-n}/s^4 and we take $n>2$. Furthermore, we fix the value of the forcing frequency at

$$\frac{\bar{\omega}_1}{2\Omega}=\frac{1}{\sqrt{3}},\tag{3.67}$$

which on the one hand allows for the resonant amplification of the EMF when $X=1/\sqrt{3}$, cf (3.54), and on the other allows to make the α effect isotropic at the leading order in the limit of weak viscosity $\nu k^2 \ll \Omega$[1],

$$\begin{aligned}
\alpha_{ij} &\approx \delta_{ij}2\pi\frac{\Delta\bar{\omega}}{\bar{\omega}_1}\frac{\mathscr{F}^2}{16\Omega^3}\int k^{3-n}\mathrm{d}k\int_0^1 \mathrm{d}X\frac{X(1-X^2)}{\left(\frac{1}{3}-X^2\right)^2+\frac{\nu^2k^4}{2\Omega^2}\left(\frac{1}{3}+X^2\right)}\\
&+\delta_{i3}\delta_{j3}2\pi\frac{\Delta\bar{\omega}}{\bar{\omega}_1}\frac{\mathscr{F}^2}{16\Omega^3}\int k^{3-n}\mathrm{d}k\int_0^1 \mathrm{d}X\frac{X(3X^2-1)}{(\varpi^2-X^2)^2+\frac{\nu^2k^4}{2\Omega^2}(\varpi^2+X^2)},\\
&\approx \delta_{ij}\frac{\Delta\bar{\omega}}{\bar{\omega}_1}\frac{\pi^2\mathscr{F}^2}{8\sqrt{3}\Omega^2\nu}\int k^{1-n}\mathrm{d}k.
\end{aligned}\tag{3.68}$$

The last integral, according to our assumptions (2.14a and b), is taken over the range of wave vectors between $K_e=2\pi/\ell_e$ and the diffusive cut-off K_ν or K_η, whichever is smaller; recall that ℓ_e denotes the length scale of the most energetic turbulent eddies. Since typically $K_e \ll K_\nu, K_\eta$, at the leading order one obtains

$$\alpha_{ij}=\alpha\delta_{ij},\quad \text{where}\quad \alpha\approx\frac{\pi^2}{16(n-2)}\frac{\Delta\bar{\omega}\mathscr{F}^2}{\Omega^3\nu K_e^{n-2}}.\tag{3.69}$$

[1] Of course the assumption of isotropy is not necessary; if $\alpha_{zz}\neq\alpha_{xx}=\alpha_{yy}$ but $\alpha_{zz}\alpha_{xx}>0$, magnetic energy growth still occurs, as described for the case of constant α_{ij} by [18].

Therefore, in this simplified picture of weak turbulence we have $\boldsymbol{\mathcal{E}} = \alpha\langle\mathbf{B}\rangle$, and the mean-field equation (2.11b) with $\eta = 0$ and $\langle\mathbf{U}\rangle = 0$ becomes

$$\frac{\partial\langle\mathbf{B}\rangle}{\partial t} = \alpha\, \nabla \times \langle\mathbf{B}\rangle. \tag{3.70}$$

This equation admits growing modes with force-free structure $\nabla \times \langle\mathbf{B}\rangle = \kappa\langle\mathbf{B}\rangle^2$, where $\kappa \ll K_e$ (to satisfy the scale separation assumed in the mean-field theory). Then (3.70) becomes

$$\frac{\partial\langle\mathbf{B}\rangle}{\partial t} = \kappa\alpha\, \langle\mathbf{B}\rangle, \tag{3.71}$$

with solution

$$\langle\mathbf{B}\rangle(\mathbf{x},\, t) = \langle\mathbf{B}\rangle(\mathbf{x},\, 0)\exp\left[\frac{\pi}{32}\frac{\kappa\Delta\bar{\omega}\mathcal{F}^2\ell_e}{\Omega^3\nu}t\right], \tag{3.72}$$

where we have chosen $n = 3$ (equal to the spatial dimension), which indicates an exponential growth of the mean field at the initial stage, with a growth rate proportional to $\Delta\bar{\omega}\ell_e\Omega^{-3}\nu^{-1}$. Consideration of a force-free mode is important, e.g., in astrophysical context of the early Universe. With negligible magnetic diffusion such a force-free mode when amplified by an α-effect can remain undamped, since there is no mechanism of energy transfer from magnetic to kinetic energy and thus no mechanism of magnetic energy loss.

The simple calculation of the α-effect induced by interactions between distinct waves in the presence of an oscillatory forcing in a highly conducting plasma allowed us to demonstrate a new 'fast' dynamo mechanism (with finite field amplification rate independent of η), that is operating even in the absence of fluid's resistivity. The mechanism of mean EMF generation from interactions of waves with distinct but close frequencies (thus the term beating waves) was identified within the regime of weak MHD turbulence (cf [19] and [3]). Such an EMF oscillates in time at a very small frequency equal to the difference between frequencies of the waves, i.e. frequencies of the distinct oscillatory components of the forcing. This time dependence is very slow, when the frequencies are close, and it was shown here, that it can be neglected at least for some initial period of evolution, $t \ll \Delta\bar{\omega}^{-1}$, and the EMF leads to amplification of a weak seed field, i.e. dynamo generation of a large-scale magnetic field. Moreover, as mentioned at the beginning of this section the assumed phase shift of $\pi/2$ between the two oscillatory components of the forcing in the form $\hat{\mathbf{f}}_1 = \mathrm{i}\hat{\mathbf{f}}_2,\, \hat{\mathbf{f}}_2 \in \mathbb{R}$, allows for a straightforward comparison between the diffusive contribution to the EMF (3.56) and the non-equilibrium one (3.51); both contributions for a given wave vector $\mathbf{k}$ have the same signs and differ only by the factor of $\Delta\bar{\omega}/\eta k^2$.

It is hypothesized, that the studied and described mechanism might be relevant to dynamo generation of magnetic fields in extremely highly conducting interstellar or

[2] Which ensures $\langle\mathbf{J}\rangle \times \langle\mathbf{B}\rangle = 0$.

intergalactic plasma in the early Universe when the galaxies were formed. It is supposed here, that at that early stage the turbulence is weak, i.e. dominated by terms linear in the turbulent fluctuations. Because of the extremely high conductivity in such systems, giving characteristic diffusion times many orders of magnitude greater than the age of the Universe, a dynamo mechanism which does not require a resistive phase shift between $\mathbf{u}'$ and $\mathbf{b}'$, which in this case would have to be extremely small, is desired. In addition, the so-generated large-scale magnetic field at the early stage could persist even at the later stages of evolution, since diffusive effects are too weak to be observable. Particular attention is focused on the force-free mode of the large-scale field, for which no mechanism of magnetic energy loss exists, when the resistivity is neglected. Such a mode can survive many Universe life-times without being damped.

The obtained α-effect is quenched, as indicated by (3.50 and b), and the quenching is of the same type as in the standard diffusive mechanism, since at $|\mathbf{k} \cdot \langle\mathbf{B}\rangle| \to \infty$ the obtained mean EMF behaves like $|\mathbf{k} \cdot \langle\mathbf{B}\rangle|^{-5}$ (cf chapter 12.6 in [2]). This quenching becomes more important as the field grows in magnitude and eventually the non-linearities lead to equilibration of the field at some level. In the present case there is also another effect, which manifests at large times, namely the time oscillations of the mean EMF on the very long time-scale $1/\Delta\bar{\omega}$, which are negligible at the leading order, but eventually start to matter. When this time-scale is much larger than the inverse growth rate of the mean field, $\Omega^3\nu/\kappa\ell_e\mathscr{F}^2 \ll 1$, the oscillations start to matter only in the already nonlinear regime, when the turbulence had become strong and therefore the first-order smoothing approximation and thus the obtained EMF (3.47) are no longer valid; the time dependence of the mean EMF is then significantly altered. Moreover, the non-stationary forcing may persist only for a period of time and then the turbulence could equilibrate to a stationary state with a large-scale magnetic field built-up through the initially effective mechanism based on interactions between beating waves. The effect of strong turbulence will be considered later in chapter 4 and chapter 5 via the renormalization method. On the other hand, the effect of long-time oscillations of the α effect and the turbulent resistivity within the weak turbulence regime is described in section 3.3, where it is shown that in natural situations, when more than two oscillatory beating modes are present in the forcing, i.e. $\mathbf{f} = \sum_j \mathfrak{Re}\left\{\hat{\mathbf{f}}_j \exp\left[\mathrm{i}(\mathbf{k} \cdot \mathbf{x} - \bar{\omega}_j t)\right]\right\}$ and at least some of the frequencies are close, the evolution of the generated mean magnetic field is likely to consist of long periods of almost stable field separated by short non-magnetized periods.

The question of whether (3.72) should be called a fast dynamo or not is perhaps a matter of semantics. In the kinematic regime in which the Lorentz force is neglected, the conventional distinction between fast and slow dynamos is clear; in the dynamical regime studied here, we have found an α-effect that oscillates on a long time-scale. Starting from an infinitesimal seed field, the long-term behaviour governed by such an α-effect will then also be oscillatory, so this is not a fast dynamo in the strict sense. However, in practice, starting from a field of small but finite amplitude, the growth rate is approximately constant and independent of η at least until nonlinear effects prevail and cause saturation, at which stage the fast/slow distinction is no longer relevant. Perhaps we may simply describe this as 'time-limited fast-dynamo' action.

3.3 The effect of long-time oscillations of the α-effect and the turbulent magnetic diffusivity on magnetic energy evolution

In this section we will obtain the expressions for the α-effect and turbulent magnetic diffusivity in weak, isotropic turbulence in the absence of background rotation

$$\mathbf{\Omega} = 0; \tag{3.73}$$

the helicity will be introduced to the flow by chiral isotropic stirring force $\mathbf{f}$. Therefore, we consider the equations (2.11a–c) for the large-scale fields and linearized equations (2.13a–c) with $\mathbf{\Omega} = 0$, i.e.

$$\begin{aligned}&\frac{\partial \mathbf{u}'}{\partial t} - \nu\nabla^2\mathbf{u}' + 2\mathbf{\Omega}\times\mathbf{u}' + (\langle\mathbf{U}\rangle\cdot\nabla)\mathbf{u}' - (\langle\mathbf{B}\rangle\cdot\nabla)\mathbf{b}' + \nabla\Pi' \\ &= \mathbf{f} - (\mathbf{u}'\cdot\nabla)\langle\mathbf{U}\rangle + (\mathbf{b}'\cdot\nabla)\langle\mathbf{B}\rangle,\end{aligned} \tag{3.74a}$$

$$\frac{\partial \mathbf{b}'}{\partial t} - \eta\nabla^2\mathbf{b}' + (\langle\mathbf{U}\rangle\cdot\nabla)\mathbf{b}' - (\langle\mathbf{B}\rangle\cdot\nabla)\mathbf{u}' = -(\mathbf{u}'\cdot\nabla)\langle\mathbf{B}\rangle + (\mathbf{b}'\cdot\nabla)\langle\mathbf{U}\rangle, \tag{3.74b}$$

$$\nabla\cdot\mathbf{b}' = 0, \quad \nabla\cdot\mathbf{u}' = 0, \tag{3.74c}$$

describing the evolution of turbulent fluctuations in weak turbulence. However, in contrast to the previous section we now retain the gradients of the mean fields on the right hand sides of the latter equations, which will be treated in a perturbational manner in order to calculate the turbulent diffusion and obtain a more complete picture of the large-scale dynamo process in weak, isotropic turbulence. Introducing

$$G_{ij} = \frac{\partial\langle U\rangle_i}{\partial x_j}, \qquad \Gamma_{ij} = \frac{\partial\langle B\rangle_i}{\partial x_j}, \tag{3.75}$$

taking the Fourier transforms of (3.74a–c) and eliminating the pressure from the Fourier transformed velocity equation with the use of the projection operator

$$P_{ij}(\mathbf{k}) = \delta_{ij} - \frac{k_i k_j}{k^2}, \tag{3.76}$$

one obtains

$$[-\mathrm{i}(\omega - \mathbf{k}\cdot\langle\mathbf{U}\rangle) + \nu k^2]\hat{\mathbf{u}} + \mathbf{P}\cdot\mathbf{G}\cdot\hat{\mathbf{u}} - \mathrm{i}\mathbf{k}\cdot\langle\mathbf{B}\rangle\hat{\mathbf{b}} - \mathbf{P}\cdot\mathbf{\Gamma}\cdot\hat{\mathbf{b}} = \hat{\mathbf{f}} \tag{3.77a}$$

$$[-\mathrm{i}(\omega - \mathbf{k}\cdot\langle\mathbf{U}\rangle) + \eta k^2]\hat{\mathbf{b}} = \mathrm{i}\mathbf{k}\cdot\langle\mathbf{B}\rangle\hat{\mathbf{u}} - \mathbf{\Gamma}\cdot\hat{\mathbf{u}} + \mathbf{G}\cdot\hat{\mathbf{b}}, \tag{3.77b}$$

$$\mathbf{k}\cdot\hat{\mathbf{b}} = 0, \quad \mathbf{k}\cdot\hat{\mathbf{u}} = 0, \tag{3.77c}$$

where we have already used (3.77c) in projecting the velocity equation on the plane perpendicular to the wave vector $\mathbf{k}$. Let us define

$$\gamma_\nu = -\mathrm{i}\omega + \nu k^2, \qquad \gamma_\eta = -\mathrm{i}\omega + \eta k^2, \tag{3.78a}$$

$$\gamma_u = \gamma_\nu + \frac{(\mathbf{k}\cdot\langle\mathbf{B}\rangle)^2}{\gamma_\eta}. \tag{3.78b}$$

and since the mean velocity $\langle \mathbf{U} \rangle$ only creates a shift of the frequency $\omega \to \omega - \mathbf{k} \cdot \langle \mathbf{U} \rangle$, we will simply absorb it into the frequency (cf also (3.6a–c) and short discussion below). Because of the assumed scale separation between the means and the fluctuations, the gradients of means can be assumed small and treated in a perturbational manner. Hence we substitute from (3.77b) into (3.77a) and into the $\mathbf{G} \cdot \hat{\mathbf{b}}$ term back in (3.77b) and neglect higher order terms in the gradients $\mathbf{G}$ and $\boldsymbol{\Gamma}$, which yields

$$\hat{\mathbf{u}} = \frac{1}{\gamma_u}\hat{\mathbf{f}} - \frac{1}{\gamma_u}\mathbf{P} \cdot \mathbf{G} \cdot \hat{\mathbf{u}} - \frac{(\mathbf{k} \cdot \langle \mathbf{B} \rangle)^2}{\gamma_u \gamma_\eta^2}\mathbf{G} \cdot \hat{\mathbf{u}} - \frac{\mathrm{i}\mathbf{k} \cdot \langle \mathbf{B} \rangle}{\gamma_u \gamma_\eta}\boldsymbol{\Gamma} \cdot \hat{\mathbf{u}} + \mathrm{i}\frac{\mathbf{k} \cdot \langle \mathbf{B} \rangle}{\gamma_u \gamma_\eta}\mathbf{P} \cdot \boldsymbol{\Gamma} \cdot \hat{\mathbf{u}}, \tag{3.79a}$$

$$\hat{\mathbf{b}} = \mathrm{i}\frac{\mathbf{k} \cdot \langle \mathbf{B} \rangle}{\gamma_\eta}\hat{\mathbf{u}} - \frac{1}{\gamma_\eta}\boldsymbol{\Gamma} \cdot \hat{\mathbf{u}} + \mathrm{i}\frac{\mathbf{k} \cdot \langle \mathbf{B} \rangle}{\gamma_\eta^2}\mathbf{G} \cdot \hat{\mathbf{u}}, \tag{3.79b}$$

The large-scale EMF, by the use of (2.12) and (3.79b), can be expressed in the following way

$$\begin{aligned} \epsilon_{ijk}\left\langle \hat{u}_j(\omega, \mathbf{k})\hat{b}_k(\omega', \mathbf{k}')\right\rangle = {} & \mathrm{i}\frac{k_n'\langle B \rangle_n}{\gamma_\eta'}\epsilon_{ijk}\left\langle \hat{u}_j\hat{u}_k'\right\rangle - \epsilon_{ijk}\frac{\partial \langle B \rangle_k}{\partial x_p}\frac{1}{\gamma_\eta'}\left\langle \hat{u}_j\hat{u}_p'\right\rangle \\ & + \mathrm{i}\epsilon_{ijk}\langle B \rangle_n\frac{\partial \langle U \rangle_k}{\partial x_p}\frac{k_n'}{\gamma_\eta'^2}\left\langle \hat{u}_j\hat{u}_p'\right\rangle, \end{aligned} \tag{3.80}$$

where we have used a short notation $\hat{u}_j' = \hat{u}_j(\omega', \mathbf{k}')$.

3.3.1 Simple kinematic theory for homogeneous turbulence

It is a common practice to assume that the background small-scale turbulence has statistical properties, that are given beforehand and thus the situation becomes kinematic. For example, we may assume, that the turbulence is homogeneous and isotropic and thus turbulence correlation tensor takes the general form

$$\left\langle \hat{u}_i(\mathbf{k}, \omega)\hat{u}_j(\mathbf{k}', \omega')\right\rangle = \left[E(\omega, \omega', k)P_{ij}(\mathbf{k}) + \mathrm{i}H(\omega, \omega', k)\epsilon_{ijk}k_k\right]\delta(\mathbf{k} + \mathbf{k}'), \tag{3.81}$$

where the function $H(\omega, \omega', k)$ is a helical term responsible for lack of reflectional symmetry, required for the large-scale dynamo process (cf [2]). The EMF takes the form

$$\begin{aligned} & \epsilon_{ljk}\int \mathrm{d}^4q\mathrm{e}^{\mathrm{i}(\mathbf{k}\cdot\mathbf{x}-\omega t)}\int \mathrm{d}^4q'\mathrm{e}^{\mathrm{i}(\mathbf{k}'\cdot\mathbf{x}-\omega' t)}\left\langle \hat{u}_j(\omega, \mathbf{k})\hat{b}_k(\omega', \mathbf{k}')\right\rangle \\ & = \frac{8\pi}{3}\langle B \rangle_l \int \mathrm{d}k \int \mathrm{d}\omega \int \mathrm{d}\omega'\mathrm{e}^{-\mathrm{i}(\omega+\omega')t}\frac{k^4}{\gamma_\eta'}H(\omega, \omega', k) \\ & \quad - \frac{8\pi}{3}[\nabla \times \langle \mathbf{B} \rangle]_l \int \mathrm{d}k \int \mathrm{d}\omega \int \mathrm{d}\omega'\mathrm{e}^{-\mathrm{i}(\omega+\omega')t}\frac{k^2}{\gamma_\eta'}E(\omega, \omega', k) \\ & \quad - \frac{4\pi}{3}[(\nabla\langle \mathbf{U} \rangle)^T \cdot \langle \mathbf{B} \rangle]_l \int \mathrm{d}k \int \mathrm{d}\omega \int \mathrm{d}\omega'\mathrm{e}^{-\mathrm{i}(\omega+\omega')t}\frac{k^4}{\gamma_\eta'^2}H(\omega, \omega', k), \end{aligned} \tag{3.82}$$

where we have introduced a shorter four-component notation

$$\mathbf{q} = (\mathbf{k}, \omega), \qquad \int d^3k \int_{-\infty}^{\infty} d\omega(\cdot) = \int d^4q(\cdot), \tag{3.83}$$

so that e.g.

$$u_i'(\mathbf{x}, t) = \int d^4q\hat{u}_i(\mathbf{q})\mathrm{e}^{\mathrm{i}(\mathbf{k}\cdot\mathbf{x}-\omega t)}. \tag{3.84}$$

Next we consider the case of non-stationary turbulence and assume the following

$$\begin{aligned} E(\omega, \omega', k) &= \mathrm{e}(\omega^2, k)\Delta(\omega, \omega'; \tilde{\omega}, \Gamma), \\ H(\omega, \omega', k) &= \mathrm{h}(\omega^2, k)\Delta(\omega, \omega'; \tilde{\omega}, \Gamma), \end{aligned} \tag{3.85}$$

where the 'non-stationarity' function

$$\Delta(\omega, \omega'; \tilde{\omega}, \Gamma) = \delta(\omega + \omega') + \Gamma\frac{\delta(\omega + \omega' + \tilde{\omega}) - \delta(\omega + \omega' - \tilde{\omega})}{2\mathrm{i}}, \tag{3.86}$$

is chosen in such a way that the correlations in the real space have a simple sinusoidal time dependence

$$\langle u_i(\mathbf{x}, t)u_i(\mathbf{x}', t)\rangle \sim 1 + \Gamma\sin(\tilde{\omega}t). \tag{3.87}$$

Let us set for simplicity $\langle \mathbf{U}\rangle = 0$, though it is straightforward to include it, and concentrate only on the α-effect and the turbulent diffusivity. Consequently, for the EMF in the real space we get

$$\begin{aligned} &\epsilon_{ljk}\int \mathrm{d}^4q\mathrm{e}^{\mathrm{i}(\mathbf{k}\cdot\mathbf{x}-\omega t)}\int \mathrm{d}^4q'\mathrm{e}^{\mathrm{i}(\mathbf{k}'\cdot\mathbf{x}-\omega' t)}\left\langle \hat{u}_j(\omega, \mathbf{k})\hat{b}_k(\omega', \mathbf{k}')\right\rangle \\ &= \frac{8\pi}{3}\eta\langle B\rangle_l\int \mathrm{d}k\int \mathrm{d}\omega\frac{k^6\mathrm{h}(\omega^2, k)}{\left|\gamma_\eta\right|^2} \\ &\quad - \mathrm{i}\frac{4\pi\Gamma}{3}\langle B\rangle_l\int \mathrm{d}k\int \mathrm{d}\omega k^4\mathrm{h}(\omega^2, k)\left[\frac{\mathrm{e}^{\mathrm{i}\tilde{\omega}t}}{\mathrm{i}(\omega + \tilde{\omega})+\eta k^2} - \frac{\mathrm{e}^{-\mathrm{i}\tilde{\omega}t}}{\mathrm{i}(\omega - \tilde{\omega})+\eta k^2}\right] \\ &\quad - \frac{8\pi}{3}\eta[\nabla\times\langle\mathbf{B}\rangle]_l\int \mathrm{d}k\int \mathrm{d}\omega\frac{k^4\mathrm{e}(\omega^2, k)}{\left|\gamma_\eta\right|^2} \\ &\quad + \mathrm{i}\frac{4\pi\Gamma}{3}[\nabla\times\langle\mathbf{B}\rangle]_l\int \mathrm{d}k\int \mathrm{d}\omega k^2\left[\frac{\mathrm{e}(\omega^2, k)\mathrm{e}^{\mathrm{i}\tilde{\omega}t}}{\mathrm{i}(\omega + \tilde{\omega})+\eta k^2} - \frac{\mathrm{e}(\omega^2, k)\mathrm{e}^{-\mathrm{i}\tilde{\omega}t}}{\mathrm{i}(\omega - \tilde{\omega})+\eta k^2}\right] \\ &= \{\eta[I_1^{(\alpha)} + \Gamma I_2^{(\alpha)}\sin(\tilde{\omega}t)] + \Gamma\tilde{\omega}I_3^{(\alpha)}\cos(\tilde{\omega}t)\}\langle B\rangle_l \\ &\quad - \{\eta[I_1^{(\eta)} + \Gamma I_2^{(\eta)}\sin(\tilde{\omega}t)] + \Gamma\tilde{\omega}I_3^{(\eta)}\cos(\tilde{\omega}t)\}[\nabla\times\langle\mathbf{B}\rangle]_l, \end{aligned} \tag{3.88}$$

where we have made use of the fact that for any even function of ω, say $f_e(\omega)$, the integral $\int_{-\infty}^{\infty}\omega f_e(\omega)\mathrm{d}\omega = 0$ vanishes and

$$I_1^{(\alpha)} = \frac{8\pi}{3}\int \mathrm{d}k\int \mathrm{d}\omega\frac{k^6\mathrm{h}(\omega^2, k)}{\left|\gamma_\eta\right|^2}, \tag{3.89a}$$

$$I_2^{(\alpha)} = \frac{8\pi}{3}\int \mathrm{d}k \int \mathrm{d}\omega \frac{k^6 \mathrm{h}(\omega^2, k)(\tilde{\omega}^2 + \omega^2 + \eta^2 k^4)}{(\omega^2 - \tilde{\omega}^2 - \eta^2 k^4)^2 + 4\omega^2\eta^2 k^4}, \tag{3.89b}$$

$$I_3^{(\alpha)} = \frac{8\pi}{3}\int \mathrm{d}k \int \mathrm{d}\omega \frac{k^4 \mathrm{h}(\omega^2, k)(\omega^2 - \tilde{\omega}^2 - \eta^2 k^4)}{(\omega^2 - \tilde{\omega}^2 - \eta^2 k^4)^2 + 4\omega^2\eta^2 k^4}, \tag{3.89c}$$

$$I_1^{(\eta)} = \frac{8\pi}{3}\int \mathrm{d}k \int \mathrm{d}\omega \frac{k^4 \mathrm{e}(\omega^2, k)}{\left|\gamma_\eta\right|^2}, \tag{3.89d}$$

$$I_2^{(\eta)} = \frac{8\pi}{3}\int \mathrm{d}k \int \mathrm{d}\omega \frac{k^4 \mathrm{e}(\omega^2, k)(\tilde{\omega}^2 + \omega^2 + \eta^2 k^4)}{(\omega^2 - \tilde{\omega}^2 - \eta^2 k^4)^2 + 4\omega^2\eta^2 k^4}, \tag{3.89e}$$

$$I_3^{(\eta)} = \frac{8\pi}{3}\int \mathrm{d}k \int \mathrm{d}\omega \frac{k^2 \mathrm{e}(\omega^2, k)(\omega^2 - \tilde{\omega}^2 - \eta^2 k^4)}{(\omega^2 - \tilde{\omega}^2 - \eta^2 k^4)^2 + 4\omega^2\eta^2 k^4}. \tag{3.89f}$$

In other words the full EMF, including the molecular diffusion, takes the following form

$$\mathcal{E}_{\mathrm{tot}} = \bar{\alpha}\langle \mathbf{B}\rangle - \bar{\eta}\nabla \times \langle \mathbf{B}\rangle, \tag{3.90}$$

where

$$\bar{\alpha} = \eta I_1^{(\alpha)} + \frac{\eta\Gamma I_2^{(\alpha)}}{\cos\phi_\alpha}\sin\left(\tilde{\omega}t + \phi_\alpha\right), \tag{3.91a}$$

$$\bar{\eta} = \eta + \eta I_1^{(\eta)} + \frac{\eta\Gamma I_2^{(\eta)}}{\cos\phi_\eta}\sin\left(\tilde{\omega}t + \phi_\eta\right), \tag{3.91b}$$

$$\tan\phi_\alpha = \frac{\tilde{\omega} I_3^{(\alpha)}}{\eta I_2^{(\alpha)}}, \qquad \tan\phi_\eta = \frac{\tilde{\omega} I_3^{(\eta)}}{\eta I_2^{(\eta)}}. \tag{3.91c}$$

Clearly, there is a significant phase shift between the $\bar{\alpha}$ coefficient and the turbulent diffusion $\bar{\eta}$.

Natural turbulence (such as e.g. in the Earth's liquid core), is obviously bound to be non-stationary, including propagation of different types of waves, such as e.g. the inertial waves (cf [20]), MAC waves (cf [21, 22]) or the magnetic Rossby waves (cf [23]), likely forming wave packets. This wave field interacts nonlinearly to form the turbulent EMF and, in particular, interactions of waves with distinct but close frequencies contribute. For simplicity we will neglect higher order interactions and include two-wave interactions only, so that in the simplest case when only two waves with distinct frequencies but the same wave vector are present (homogeneity but non-stationarity), the turbulent point correlations at time t can be written as

$$
\begin{aligned}
&\langle \mathbf{u}(\mathbf{x},\, t)\mathbf{u}(\mathbf{x},\, t)\rangle \\
&= \langle [\hat{\mathbf{u}}_1 e^{i(\omega_1 t+\mathbf{k}\cdot\mathbf{x})} + \hat{\mathbf{u}}_2 e^{i(\omega_2 t+\mathbf{k}\cdot\mathbf{x})} + c.\ c.][\hat{\mathbf{u}}_1^* e^{-i(\omega_1 t+\mathbf{k}\cdot\mathbf{x})} + \hat{\mathbf{u}}_2^* e^{-i(\omega_2 t+\mathbf{k}\cdot\mathbf{x})} + c.\ c.]\rangle \\
&= \langle \hat{\mathbf{u}}_1\hat{\mathbf{u}}_2^* e^{i\tilde{\omega}t}\rangle + \langle \hat{\mathbf{u}}_2\hat{\mathbf{u}}_1^* e^{-i\tilde{\omega}t}\rangle + \langle \hat{\mathbf{u}}_1\hat{\mathbf{u}}_1^*\rangle + \langle \hat{\mathbf{u}}_2\hat{\mathbf{u}}_2^*\rangle,
\end{aligned} \tag{3.92}
$$

where

$$
\tilde{\omega} = \Delta\omega = \omega_1 - \omega_2, \tag{3.93}
$$

$c.\ c.$ means the complex conjugate and we have assumed that the means of quantities oscillating with frequencies ω_1, ω_2 and $\omega_1 + \omega_2$, which correspond to the short fluctuational timescales can be assumed to vanish (without loss of generality we may assume $\omega_1 > 0$, $\omega_2 > 0$). When the frequency difference $\tilde{\omega}$ is small the first two terms in the second line of (3.92) cannot be neglected and lead to long time variations of the turbulence correlation tensor. This is in keeping with the chosen form of the 'non-stationarity' function (3.86), which leads to correlations in the form (3.87) corresponding to the form based on two-wave interactions.

In reality even such two-wave interactions lead to more than one (say N) $\tilde{\omega}$-modes, because the turbulent wave field consists of a number of distinct waves so that more generally

$$
\langle u_i(\mathbf{x},\, t)u_i(\mathbf{x}',\, t)\rangle \sim 1 + \sum_{n=1}^{N}\Gamma_n \sin(\tilde{\omega}_n t + \psi_n), \tag{3.94}
$$

where the phase shift ψ_n accounts for both, the sine-type and cosine-type time dependence of the correlation tensor.

3.3.2 Dynamic theory

Instead of assuming some given form of the turbulence correlation tensor we can use the equation (3.79) to express it in the following form

$$
\begin{aligned}
\langle \hat{u}_i(\mathbf{q})\hat{u}_j(\mathbf{q}')\rangle &= \frac{1}{\gamma_u\gamma_u'}\langle \hat{f}_i\hat{f}_j'\rangle \\
&- \left[\frac{1}{\gamma_u\gamma_u'^2}P_{js}'G_{sp} + \frac{(\mathbf{k}'\cdot\langle\mathbf{B}\rangle)^2}{\gamma_u\gamma_u'^2\gamma_\eta'^2}G_{jp} + i\frac{\mathbf{k}'\cdot\langle\mathbf{B}\rangle}{\gamma_u\gamma_u'^2\gamma_\eta'}\frac{k_j'k_s'}{k'^2}\Gamma_{sp}\right]\langle \hat{f}_i\hat{f}_p'\rangle \\
&- \left[\frac{1}{\gamma_u^2\gamma_u'}P_{is}G_{sp} + \frac{(\mathbf{k}\cdot\langle\mathbf{B}\rangle)^2}{\gamma_u^2\gamma_\eta^2\gamma_u'}G_{ip} + i\frac{\mathbf{k}\cdot\langle\mathbf{B}\rangle}{\gamma_u^2\gamma_\eta\gamma_u'}\frac{k_ik_s}{k^2}\Gamma_{sp}\right]\langle \hat{f}_p\hat{f}_j'\rangle,
\end{aligned} \tag{3.95}
$$

where higher order terms in the gradients $\mathbf{G}$ and $\boldsymbol{\Gamma}$ have been neglected and we have used the short four-component notation $\mathbf{q} = (\omega, \mathbf{k})$. The EMF can then be calculated with the use of (3.80) and (3.95) which yields

$$
\begin{aligned}
\epsilon_{ljk}\left\langle \hat{u}_j(\mathbf{q})\hat{b}_k(\mathbf{q}')\right\rangle = {} & \mathrm{i}\frac{k_n'\langle B\rangle_n}{\gamma_\eta'}\epsilon_{ljk}\frac{1}{\gamma_u\gamma_u'}\left\langle \hat{f}_j\hat{f}_k'\right\rangle \\
& + \langle B\rangle_n\langle B\rangle_r\epsilon_{ljk}\frac{1}{\gamma_u\gamma_u'^2\gamma_\eta'^2}\frac{k_n'k_r'k_k'k_s'}{k'^2}\frac{\partial\langle B\rangle_s}{\partial x_p}\left\langle \hat{f}_j\hat{f}_p'\right\rangle \\
& + \langle B\rangle_n\langle B\rangle_r\epsilon_{ljk}\frac{1}{\gamma_u^2\gamma_\eta\gamma_\eta'\gamma_u'}\frac{k_n'k_rk_jk_s}{k^2}\frac{\partial\langle B\rangle_s}{\partial x_p}\left\langle \hat{f}_p\hat{f}_k'\right\rangle \\
& - \epsilon_{ljk}\frac{\partial\langle B\rangle_k}{\partial x_p}\frac{1}{\gamma_u\gamma_u'\gamma_\eta'}\left\langle \hat{f}_j\hat{f}_p'\right\rangle \\
& - \mathrm{i}\frac{k_n'\langle B\rangle_n}{\gamma_\eta'}\epsilon_{ljk}\left[\frac{1}{\gamma_u\gamma_u'^2}P_{ks}'\frac{\partial\langle U\rangle_s}{\partial x_p} + \frac{(\mathbf{k}'\cdot\langle\mathbf{B}\rangle)^2}{\gamma_u\gamma_u'^2\gamma_\eta'^2}\frac{\partial\langle U\rangle_k}{\partial x_p}\right]\left\langle \hat{f}_j\hat{f}_p'\right\rangle \\
& - \mathrm{i}\frac{k_n'\langle B\rangle_n}{\gamma_\eta'}\epsilon_{ljk}\left[\frac{1}{\gamma_u^2\gamma_u'}P_{js}\frac{\partial\langle U\rangle_s}{\partial x_p} + \frac{(\mathbf{k}\cdot\langle\mathbf{B}\rangle)^2}{\gamma_u^2\gamma_\eta^2\gamma_u'}\frac{\partial\langle U\rangle_j}{\partial x_p}\right]\left\langle \hat{f}_p\hat{f}_k'\right\rangle \\
& + \mathrm{i}\epsilon_{ljk}\langle B\rangle_n\frac{\partial\langle U\rangle_k}{\partial x_p}\frac{k_n'}{\gamma_u\gamma_u'\gamma_\eta'^2}\left\langle \hat{f}_j\hat{f}_p'\right\rangle .
\end{aligned}
\tag{3.96}
$$

On assuming the force correlations in a statistically homogeneous and isotropic form

$$
\left\langle \hat{f}_i(\mathbf{k},\,\omega)\hat{f}_j(\mathbf{k}',\,\omega')\right\rangle = \left[F_0(\omega,\,\omega',\,k)P_{ij}(\mathbf{k}) + \mathrm{i}F_1(\omega,\,\omega',\,k)\epsilon_{ijk}k_k\right]\delta(\mathbf{k}+\mathbf{k}'),
\tag{3.97}
$$

we return to the real space to get for the EMF

$$
\begin{aligned}
& \epsilon_{ljk}\int \mathrm{d}^4q\,\mathrm{e}^{\mathrm{i}(\mathbf{k}\cdot\mathbf{x}-\omega t)}\int \mathrm{d}^4q'\,\mathrm{e}^{\mathrm{i}(\mathbf{k}'\cdot\mathbf{x}-\omega' t)}\left\langle \hat{u}_j(\omega,\,\mathbf{k})\hat{b}_k(\omega',\,\mathbf{k}')\right\rangle \\
& = 2\langle B\rangle_n\int \mathrm{d}^4q\int \mathrm{d}\omega'\mathrm{e}^{-\mathrm{i}(\omega+\omega')t}\frac{k_nk_l}{\gamma_u\gamma_u'\gamma_\eta'}F_1 \\
& \quad - \epsilon_{ljk}\frac{\partial\langle B\rangle_k}{\partial x_p}\int \mathrm{d}^4q\int \mathrm{d}\omega'\mathrm{e}^{-\mathrm{i}(\omega+\omega')t}\frac{1}{\gamma_u\gamma_u'\gamma_\eta'}F_0P_{jp}(\mathbf{k}) \\
& \quad + \langle B\rangle_n\langle B\rangle_r\frac{\partial\langle B\rangle_s}{\partial x_j}\epsilon_{ljk}\int \mathrm{d}^4q\int \mathrm{d}\omega'\mathrm{e}^{-\mathrm{i}(\omega+\omega')t}\frac{1}{\gamma_u\gamma_u'^2\gamma_\eta'^2}\frac{k_nk_rk_kk_s}{k^2}F_0 \\
& \quad + \langle B\rangle_n\langle B\rangle_r\frac{\partial\langle B\rangle_s}{\partial x_j}\epsilon_{ljk}\int \mathrm{d}^4q\int \mathrm{d}\omega'\mathrm{e}^{-\mathrm{i}(\omega+\omega')t}\frac{1}{\gamma_u^2\gamma_u'\gamma_\eta\gamma_\eta'}\frac{k_nk_rk_kk_s}{k^2}F_0 \\
& \quad + \langle B\rangle_n\frac{\partial\langle U\rangle_s}{\partial x_k}\int \mathrm{d}^4q\int \mathrm{d}\omega'\mathrm{e}^{-\mathrm{i}(\omega+\omega')t}\left(\frac{1}{\gamma_u'} + \frac{1}{\gamma_u}\right)\frac{k_nk_lk_kk_s}{k^2\gamma_u\gamma_u'\gamma_\eta'}F_1 \\
& \quad + \langle B\rangle_n\langle B\rangle_r\langle B\rangle_w\frac{\partial\langle U\rangle_k}{\partial x_l}\int \mathrm{d}^4q\int \mathrm{d}\omega'\mathrm{e}^{-\mathrm{i}(\omega+\omega')t}\left(\frac{1}{\gamma_u'\gamma_\eta'^2} + \frac{1}{\gamma_u\gamma_\eta^2}\right)\frac{k_nk_kk_rk_w}{\gamma_u\gamma_u'\gamma_\eta'}F_1 \\
& \quad - \langle B\rangle_n\frac{\partial\langle U\rangle_k}{\partial x_l}\int \mathrm{d}^4q\int \mathrm{d}\omega'\mathrm{e}^{-\mathrm{i}(\omega+\omega')t}\frac{k_nk_k}{\gamma_u\gamma_u'\gamma_\eta'^2}F_1 .
\end{aligned}
\tag{3.98}
$$

Next, using the following formulae

$$\int \frac{k_j}{k} f(\cos^2\theta)\mathrm{d}\mathring{\Omega} = 0, \qquad \int \frac{k_i k_j k_k}{k^3} f(\cos\theta)\mathrm{d}\mathring{\Omega} = 0, \tag{3.99a}$$

$$\int \frac{k_j k_n}{k^2} f(\cos^2\theta)\mathrm{d}\mathring{\Omega} = \pi \int_{-1}^{1} f(X^2)\left\{\delta_{jn}(1 - X^2) + \delta_{j3}\delta_{n3}(3X^2 - 1)\right\}\mathrm{d}X, \tag{3.99b}$$

$$\begin{aligned}
&\int \frac{k_i k_j k_m k_n}{k^4} f(\cos^2\theta)\mathrm{d}\mathring{\Omega} \\
&= \frac{\pi}{4}\left(\delta_{ij}\delta_{mn} + \delta_{im}\delta_{jn} + \delta_{in}\delta_{jm}\right)\int_{-1}^{1} f(X^2)(1 - X^2)^2\mathrm{d}X \\
&\quad - \frac{\pi}{4}\left(\delta_{ij}\delta_{m3}\delta_{n3} + \delta_{im}\delta_{j3}\delta_{n3} + \delta_{in}\delta_{j3}\delta_{m3} + \delta_{jm}\delta_{i3}\delta_{n3} + \delta_{jn}\delta_{i3}\delta_{m3}\right. \\
&\quad \left. + \delta_{mn}\delta_{i3}\delta_{j3}\right)\int_{-1}^{1} f(X^2)(5X^4 - 6X^2 + 1)\mathrm{d}X \\
&\quad + \frac{\pi}{4}\delta_{i3}\delta_{j3}\delta_{m3}\delta_{n3}\int_{-1}^{1} f(X^2)(35X^4 - 30X^2 + 3)\mathrm{d}X,
\end{aligned} \tag{3.99c}$$

where $\mathring{\Omega}$ denotes the solid angle and the spherical coordinates (k, θ, φ) have been used (with the substitution $X = \cos\theta$) one obtains

$$\begin{aligned}
&\epsilon_{ljk}\int \mathrm{d}^4 q e^{\mathrm{i}(\mathbf{k}\cdot\mathbf{x}-\omega t)}\int \mathrm{d}^4 q' e^{\mathrm{i}(\mathbf{k}'\cdot\mathbf{x}-\omega' t)}\left\langle \hat{u}_j(\omega, \mathbf{k})\hat{b}_k(\omega', \mathbf{k}')\right\rangle \\
&= \mathcal{I}_1(\nu, \eta, \langle B\rangle^2)\langle B\rangle_l - [\mathcal{I}_2(\nu, \eta, \langle B\rangle^2) + \langle B\rangle^2\mathcal{I}_3(\nu, \eta, \langle B\rangle^2)][\nabla \times \langle\mathbf{B}\rangle]_l \\
&- \mathcal{I}_4(\nu, \eta, \langle B\rangle^2)[\nabla\langle B\rangle^2 \times \langle\mathbf{B}\rangle]_l \\
&+ \left\{\mathcal{I}_6(\nu, \eta, \langle B\rangle^2) - \mathcal{I}_5(\nu, \eta, \langle B\rangle^2) + \langle B\rangle^2\mathcal{I}_8(\nu, \eta, \langle B\rangle^2)\right\}[(\nabla\langle\mathbf{U}\rangle)^{(s)}\cdot\langle\mathbf{B}\rangle]_l \\
&+ \mathcal{I}_7(\nu, \eta, \langle B\rangle^2)\frac{\langle\mathbf{B}\rangle\cdot(\nabla\langle\mathbf{U}\rangle)^{(s)}\cdot\langle\mathbf{B}\rangle}{\langle B\rangle^2}\langle B\rangle_l \\
&- \left\{\mathcal{I}_8(\nu, \eta, \langle B\rangle^2)\langle B\rangle^2 - \mathcal{I}_5(\nu, \eta, \langle B\rangle^2)\right\}[(\nabla \times \langle\mathbf{U}\rangle) \times \langle\mathbf{B}\rangle]_l,
\end{aligned} \tag{3.100}$$

where $(\nabla\langle\mathbf{U}\rangle)^{(s)}$ is the symmetric part of the mean velocity gradient tensor. The integrals $\mathcal{I}_j(\nu, \eta, \langle B\rangle^2)$ are defined as

$$\mathcal{I}_1(\nu, \eta, \langle B\rangle^2) = 4\pi \iiiint e^{-\mathrm{i}(\omega+\omega')t}\frac{k^4 X^2 F_1(\omega, \omega', k)}{\gamma_u\gamma_u'\gamma_\eta'}, \tag{3.101a}$$

$$\mathcal{I}_2(\nu, \eta, \langle B\rangle^2) = \pi \iiiint e^{-\mathrm{i}(\omega+\omega')t}\frac{k^2(1 + X^2)F_0}{\gamma_u\gamma_u'\gamma_\eta'}, \tag{3.101b}$$

$$\mathcal{I}_3(\nu, \eta, \langle B\rangle^2) = \frac{\pi}{2}\iiiint \mathrm{e}^{-\mathrm{i}(\omega+\omega')t}\frac{k^4F_0}{\gamma_u\gamma_u'\gamma_\eta'}\left(\frac{1}{\gamma_u'\gamma_\eta'} + \frac{1}{\gamma_u\gamma_\eta}\right)(-13X^4 + 10X^2 - 1), \tag{3.101c}$$

$$\mathcal{I}_4(\nu, \eta, \langle B\rangle^2) = \frac{\pi}{4}\iiiint \mathrm{e}^{-\mathrm{i}(\omega+\omega')t}\frac{k^4F_0}{\gamma_u\gamma_u'\gamma_\eta'}\left(\frac{1}{\gamma_u'\gamma_\eta'} + \frac{1}{\gamma_u\gamma_\eta}\right)(9X^4 - 10X^2 + 1), \tag{3.101d}$$

$$\mathcal{I}_5(\nu, \eta, \langle B\rangle^2) = 2\pi\iiiint \mathrm{e}^{-\mathrm{i}(\omega+\omega')t}\frac{k^4X^2F_1}{\gamma_u\gamma_u'\gamma_\eta'^2}, \tag{3.101e}$$

$$\mathcal{I}_6(\nu, \eta, \langle B\rangle^2) = 2\pi\iiiint \mathrm{e}^{-\mathrm{i}(\omega+\omega')t}\frac{k^4F_1}{\gamma_u\gamma_u'\gamma_\eta'}\left(\frac{1}{\gamma_u'} + \frac{1}{\gamma_u}\right)X^2(1 - X^2), \tag{3.101f}$$

$$\mathcal{I}_7(\nu, \eta, \langle B\rangle^2) = \pi\iiiint \mathrm{e}^{-\mathrm{i}(\omega+\omega')t}\frac{k^4F_1}{\gamma_u\gamma_u'\gamma_\eta'}\left(\frac{1}{\gamma_u'} + \frac{1}{\gamma_u}\right)X^2(5X^2 - 3), \tag{3.101g}$$

$$\mathcal{I}_8(\nu, \eta, \langle B\rangle^2) = 2\pi\iiiint \mathrm{e}^{-\mathrm{i}(\omega+\omega')t}\frac{k^6X^4F_1}{\gamma_u\gamma_u'\gamma_\eta'}\left(\frac{1}{\gamma_u'\gamma_\eta'^2} + \frac{1}{\gamma_u\gamma_\eta^2}\right), \tag{3.101h}$$

where

$$\iiiint = \int \mathrm{d}k \int_{-1}^{1} \mathrm{d}X \int_{-\infty}^{\infty} \mathrm{d}\omega \int_{-\infty}^{\infty} \mathrm{d}\omega'. \tag{3.101i}$$

Using vector identity $(\nabla \times \langle \mathbf{U}\rangle) \times \langle \mathbf{B}\rangle = 2(\nabla\langle \mathbf{U}\rangle)^{(a)} \cdot \langle \mathbf{B}\rangle$, where $(\nabla\langle \mathbf{U}\rangle)^{(a)}$ denotes the antisymmetric part of the mean velocity gradient tensor we can easily see that all the terms involving the large-scale shear, i.e. the mean velocity gradients contribute to the α-effect and make it anisotropic. It is also worth noting, that the third and last terms in (3.100) contribute to an effective large-scale drift velocity of the mean magnetic field

$$\begin{aligned} &- \mathcal{I}_4(\nu, \eta, \langle B\rangle^2)\nabla\langle B\rangle^2 \times \langle \mathbf{B}\rangle \\ &- \left\{\mathcal{I}_8(\nu, \eta, \langle B\rangle^2)\langle B\rangle^2 - \mathcal{I}_5(\nu, \eta, \langle B\rangle^2)\right\}(\nabla \times \langle \mathbf{U}\rangle) \times \langle \mathbf{B}\rangle = \mathbf{v}_{\mathrm{drift}} \times \langle \mathbf{B}\rangle, \end{aligned} \tag{3.102}$$

and both belong to the anisotropic α-effect. Although it is straightforward to include the anisotropic terms in the calculations, for clarity we will exclude the anisotropy resulting from mean-field gradients, i.e. we set $\langle \mathbf{U}\rangle = 0$ and neglect the effect of the drift velocity. In such a way we concentrate only on the isotropic part of the α-effect and the effects of turbulent diffusion, also isotropic in this case. Hence the mean EMF is reduced to

$$\mathcal{E}_l = \epsilon_{ljk}\int d^4q e^{i(\mathbf{k}\cdot\mathbf{x}-\omega t)}\int d^4q' e^{i(\mathbf{k}'\cdot\mathbf{x}-\omega' t)}\left\langle \hat{u}_j(\omega, \mathbf{k})\hat{b}_k(\omega', \mathbf{k}')\right\rangle$$
$$= \mathcal{I}_1(\nu, \eta, \langle B\rangle^2)\langle B\rangle_l - [\mathcal{I}_2(\nu, \eta, \langle B\rangle^2) + \langle B\rangle^2\mathcal{I}_3(\nu, \eta, \langle B\rangle^2)][\nabla\times\langle\mathbf{B}\rangle]_l, \tag{3.103}$$

where the integrals $\mathcal{I}_1$, $\mathcal{I}_2$ and $\mathcal{I}_3$ are provided in (3.101a–c).

3.3.2.1 Stationary turbulence

First let us explicitly demonstrate the simplest result obtained for stationary turbulence, when

$$F_0(\omega, \omega', k) = \mathcal{F}_0(\omega^2, k)\delta(\omega+\omega'), \quad F_1(\omega, \omega', k) = \mathcal{F}_1(\omega^2, k)\delta(\omega+\omega'), \tag{3.104}$$

i.e. the force correlations are given by

$$\left\langle \hat{f}_i(\mathbf{q})\hat{f}_j(\mathbf{q}')\right\rangle = \left[\mathcal{F}_0(\omega^2, k)P_{ij}(\mathbf{k}) + i\mathcal{F}_1(\omega^2, k)\epsilon_{ijk}k_k\right]\delta(\mathbf{k}+\mathbf{k}')\delta(\omega+\omega'). \tag{3.105}$$

In such a case the integrals in (3.103) can be calculated to yield

$$\mathcal{I}_1^{(st)}(\nu, \eta, \langle B\rangle^2) = 4\pi\eta\iiint\frac{k^6X^2\mathcal{F}_1(\omega^2, k)}{|\gamma_u|^2|\gamma_\eta|^2}, \tag{3.106a}$$

$$\mathcal{I}_2^{(st)}(\nu, \eta, \langle B\rangle^2) = \pi\eta\iiint\frac{k^4(1+X^2)\mathcal{F}_0(\omega^2, k)}{|\gamma_u|^2|\gamma_\eta|^2}, \tag{3.106b}$$

$$\mathcal{I}_3^{(st)}(\nu, \eta, \langle B\rangle^2)$$
$$= \pi\eta\iiint\frac{k^6\mathcal{F}_0(\omega^2, k)}{|\gamma_u|^4|\gamma_\eta|^4}(\omega^2 - \nu\eta k^4 - (\mathbf{k}\cdot\langle\mathbf{B}\rangle)^2)(13X^4 - 10X^2 + 1), \tag{3.106c}$$

where

$$\iiint = \int dk\int_{-1}^{1} dX\int_{-\infty}^{\infty} d\omega. \tag{3.107}$$

Therefore, the induction equation takes the following form

$$\frac{\partial\langle\mathbf{B}\rangle}{\partial t} = \nabla\times(\bar{\alpha}_{st}\langle\mathbf{B}\rangle) - \nabla\times(\bar{\eta}_{st}\nabla\times\langle\mathbf{B}\rangle), \tag{3.108}$$

with

$$\bar{\alpha}_{st} = 4\pi\eta\iiint\frac{k^6X^2\mathcal{F}_1(\omega^2, k)}{|\gamma_u|^2|\gamma_\eta|^2}, \tag{3.109a}$$

$$\bar{\eta}_{st} = \eta\left\{1 + \pi \iiint \frac{k^4\mathcal{F}_0(\omega^2, k)}{|\gamma_u|^2|\gamma_\eta|^2}\left[1 + X^2 \right.\right.$$
$$\left.\left. + \langle B\rangle^2 \frac{k^2(\omega^2 - \nu\eta k^4 - k^2\langle B\rangle^2 X^2)(13X^4 - 10X^2 + 1)}{|\gamma_u|^2|\gamma_\eta|^2}\right]\right\}, \tag{3.109b}$$

where we have made use of the fact, that for any even function of ω, say $f_e(\omega)$, the integral $\int_{-\infty}^{\infty} \omega f_e(\omega)\mathrm{d}\omega = 0$ vanishes.

3.3.2.2 Non-stationary turbulence

The turbulent wave field is composed of the distinct fundamental modes, i.e. the Lehnert waves. In very complex systems such as the Earth's core, in addition to the Lorentz and Coriolis forces the turbulent flow is strongly influenced also by the buoyancy forces and geometrical effects and the wave field involves also the so-called MAC waves (*M*agnetic-*A*rchimedean-*C*oriolis) and the magnetic Rossby waves, cf [21, 22] and [23]. These waves are described by different dispersion relations which implies existence of a large number of waves with distinct frequencies of oscillation. As argued at the end of section 3.3.1 their interactions lead to time dependent turbulent correlation tensor. We consider the simplest case of only two-wave interactions and introduce a simple model by postulating that the turbulence is forced by the following isotropic, homogeneous but non-stationary forcing (cf (3.86))

$$F_0(\omega, \omega', k) = \mathcal{F}_0(\omega^2, k)\Delta(\omega, \omega'; \tilde{\omega}, \Gamma), \tag{3.110a}$$

$$F_1(\omega, \omega', k) = \mathcal{F}_1(\omega^2, k)\Delta(\omega, \omega'; \tilde{\omega}, \Gamma), \tag{3.110b}$$

i.e.

$$\left\langle \hat{f}_i(\mathbf{q})\hat{f}_j(\mathbf{q}')\right\rangle = \left[\mathcal{F}_0(\omega^2, k)P_{ij}(\mathbf{k}) + \mathrm{i}\mathcal{F}_1(\omega^2, k)\epsilon_{ijk}k_k\right]\delta(\mathbf{k} + \mathbf{k}')\Delta(\omega, \omega'; \tilde{\omega}, \Gamma). \tag{3.111}$$

In the above the function $\Delta(\omega, \omega'; \tilde{\omega}, \Gamma)$ is the 'non-stationarity' function given in (3.86), so that indeed the two-wave interaction form

$$\left\langle f_i(\mathbf{x}, t)f_i(\mathbf{x}', t)\right\rangle \sim 1 + \Gamma\sin(\tilde{\omega}t) \tag{3.112}$$

is preserved[3]. The parameter $\tilde{\omega}$ has the physical interpretation of frequency shift between distinct waves in the turbulent wave field.

The integrals (3.101–c) take the following form

$$\mathcal{I}_1(\nu, \eta, \langle B\rangle^2) = \mathcal{I}_1^{(st)}(\nu, \eta, \langle B\rangle^2) - \mathrm{i}2\pi\Gamma \iiint \frac{k^4X^2\mathcal{F}_1(\omega^2, k)}{\gamma_u}\mathcal{W}(\omega; \tilde{\omega}, \nu, \eta), \tag{3.113a}$$

[3] Of course equally well one could choose this to be a cosine-type time dependence.

$$\mathcal{I}_2(\nu, \eta, \langle B\rangle^2) = \mathcal{I}_2^{(st)}(\nu, \eta, \langle B\rangle^2) - \mathrm{i}\frac{\pi}{2}\Gamma \iiint \frac{k^2(1+X^2)\mathcal{F}_0(\omega^2, k)}{\gamma_u}\mathcal{W}(\omega; \tilde{\omega}, \nu, \eta), \tag{3.113b}$$

$$\begin{aligned}\mathcal{I}_3(\nu, \eta, \langle B\rangle^2) = {}& \mathcal{I}_3^{(st)}(\nu, \eta, \langle B\rangle^2) \\ & - \mathrm{i}\frac{\pi}{4}\Gamma \iiint \frac{k^4\mathcal{F}_0(\omega^2, k)}{\gamma_u^2\gamma_\eta}(-13X^4 + 10X^2 - 1)\mathcal{W}(\omega; \tilde{\omega}, \nu, \eta) \\ & - \mathrm{i}\frac{\pi}{4}\Gamma \iiint \frac{k^4\mathcal{F}_0(\omega^2, k)}{\gamma_u}(-13X^4 + 10X^2 - 1)\overline{\mathcal{W}}(\omega; \tilde{\omega}, \nu, \eta),\end{aligned} \tag{3.113c}$$

where

$$\mathcal{W}(\omega; \tilde{\omega}, \nu, \eta) = \frac{e^{\mathrm{i}\tilde{\omega}t}}{\gamma_u(-\omega-\tilde{\omega})\gamma_\eta(-\omega-\tilde{\omega})} - \frac{e^{-\mathrm{i}\tilde{\omega}t}}{\gamma_u(-\omega+\tilde{\omega})\gamma_\eta(-\omega+\tilde{\omega})}, \tag{3.114a}$$

$$\overline{\mathcal{W}}(\omega; \tilde{\omega}, \nu, \eta) = \frac{e^{\mathrm{i}\tilde{\omega}t}}{\gamma_u^2(-\omega-\tilde{\omega})\gamma_\eta^2(-\omega-\tilde{\omega})} - \frac{e^{-\mathrm{i}\tilde{\omega}t}}{\gamma_u^2(-\omega+\tilde{\omega})\gamma_\eta^2(-\omega+\tilde{\omega})}, \tag{3.114b}$$

and $\mathcal{I}_j^{(st)}(\nu, \eta, \langle B\rangle^2)$ are given in (3.106a–c). To make analytical progress we now consider the following asymptotic limit

$$\forall_k \; \tilde{\omega} \ll \nu k^2, \; \eta k^2, \tag{3.115}$$

which is in keeping with the limit of beating waves (3.49) studied in section 3.2, so that

$$\begin{aligned}&\gamma_u(-\omega-\tilde{\omega})\gamma_\eta(-\omega-\tilde{\omega}) \\ &= \left\{\nu k^2 + \eta k^2\frac{(\mathbf{k}\cdot\langle\mathbf{B}\rangle)^2}{|\gamma_\eta|^2} + \mathrm{i}(-\omega-\tilde{\omega})\left[\frac{(\mathbf{k}\cdot\langle\mathbf{B}\rangle)^2}{|\gamma_\eta|^2} - 1\right]\right\}(\eta k^2 - \mathrm{i}(-\omega-\tilde{\omega})) \\ &= \nu\eta k^4 - (\omega+\tilde{\omega})^2 + (\mathbf{k}\cdot\langle\mathbf{B}\rangle)^2 + \mathrm{i}(\omega+\tilde{\omega})(\nu+\eta)k^2 \\ &\approx \nu\eta k^4 - \omega^2 + (\mathbf{k}\cdot\langle\mathbf{B}\rangle)^2 + \mathrm{i}\omega(\nu+\eta)k^2 + \tilde{\omega}[\mathrm{i}(\nu+\eta)k^2 - 2\omega],\end{aligned} \tag{3.116}$$

$$\frac{1}{\gamma_u(-\omega-\tilde{\omega})\gamma_\eta(-\omega-\tilde{\omega})} \approx \frac{1}{\gamma_u(-\omega)\gamma_\eta(-\omega)} - \tilde{\omega}\frac{\mathrm{i}(\nu+\eta)k^2 - 2\omega}{\gamma_u^2(-\omega)\gamma_\eta^2(-\omega)}. \tag{3.117}$$

Let us recall that according to the preliminary results of section 3.2 the assumption (3.115) can make the non-equilibrium dynamo effect significantly weaker, however, it is required to obtain quantitative analytic results; moreover, as shown below even under this restrictive assumption the non-equilibrium contribution to the α-effect still becomes dominant in many cases. On introducing the following new variable

$$\varpi = \frac{\omega}{\sqrt{\nu\eta}\,k^2}, \quad \text{and defining} \quad \mathcal{M}^2(k) = \frac{\langle B\rangle^2}{\nu\eta k^2}, \tag{3.118}$$

we get

$$
\begin{aligned}
&| \gamma_u(\omega)|^2 |\gamma_\eta(\omega)|^2 \\
&= \omega^4 + \omega^2[k^4(\nu^2 + \eta^2) - 2(\mathbf{k} \cdot \langle \mathbf{B} \rangle)^2] + (\nu\eta k^4 + (\mathbf{k} \cdot \langle \mathbf{B} \rangle)^2)^2 \\
&= \nu^2\eta^2 k^8 \left\{ \varpi^4 + \varpi^2[(Pm + Pm^{-1}) - 2\mathcal{M}^2 X^2] + (1 + \mathcal{M}^2 X^2)^2 \right\},
\end{aligned}
\tag{3.119}
$$

and hence with the use of (3.111) one obtains

$$\mathcal{I}_1(\nu, \eta, \langle B \rangle^2) = \bar{\alpha}_{\rm st} + \bar{\alpha}_{\rm st}\Gamma \sin(\tilde{\omega}t) - \bar{\alpha}_{\Delta\omega}\Gamma \cos(\tilde{\omega}t) \tag{3.120a}$$

$$\mathcal{I}_2(\nu, \eta, \langle B \rangle^2) = \bar{\eta}_{1\,st} + \bar{\eta}_{1\,st}\Gamma \sin(\tilde{\omega}t) - \bar{\eta}_{1\,\Delta\omega}\Gamma \cos(\tilde{\omega}t), \tag{3.120b}$$

$$\langle B \rangle^2 \mathcal{I}_3(\nu, \eta, \langle B \rangle^2) = -\bar{\eta}_{2\,st} - \bar{\eta}_{2\,st}\Gamma \sin(\tilde{\omega}t) + \bar{\eta}_{2\,\Delta\omega}\Gamma \cos(\tilde{\omega}t), \tag{3.120c}$$

where

$$\bar{\alpha}_{\rm st} = \frac{4\pi}{\nu\sqrt{\nu\eta}} \int {\rm d}k \int_{-1}^{1} {\rm d}X \int {\rm d}\varpi \frac{\mathcal{F}_1(\varpi^2, k)X^2}{\mathcal{Z}}, \tag{3.121a}$$

$$
\begin{aligned}
\bar{\alpha}_{\Delta\omega} &= -4\pi\tilde{\omega} \int {\rm d}k \int_{-1}^{1} {\rm d}X \int {\rm d}\omega \frac{k^4 X^2 \mathcal{F}_1(\omega^2, k)}{| \gamma_u |^4 |\gamma_\eta|^4} \\
&\quad \times \left\{ 2\omega^4 + \omega^2[\nu(\nu - \eta)k^4 - 2k^2\langle B \rangle^2 X^2] - (\nu + \eta)\eta k^4[\nu\eta k^4 + k^2\langle B \rangle^2 X^2] \right\} \\
&= -\frac{4\pi\tilde{\omega}}{(\nu\eta)^{3/2}} \int \frac{{\rm d}k}{k^2} \int_{-1}^{1} {\rm d}X \int {\rm d}\varpi \frac{\mathcal{F}_1(\varpi^2, k)X^2}{\mathcal{Z}^2} \\
&\quad \times [2\varpi^4 + \varpi^2(Pm - 1 - 2\mathcal{M}^2 X^2) - (1 + Pm^{-1})(1 + \mathcal{M}^2 X^2)],
\end{aligned}
\tag{3.121b}
$$

$$
\begin{aligned}
\bar{\eta}_{1\,st} &= \pi\eta \int {\rm d}k \int_{-1}^{1} {\rm d}X \int {\rm d}\omega \frac{k^4(1 + X^2)\mathcal{F}_0(\omega^2, k)}{| \gamma_u |^2 |\gamma_\eta|^2} \\
&= \frac{\pi}{\nu\sqrt{\nu\eta}} \int \frac{{\rm d}k}{k^2} \int_{-1}^{1} {\rm d}X \int {\rm d}\varpi \frac{\mathcal{F}_0(\varpi^2, k)(1 + X^2)}{\mathcal{Z}},
\end{aligned}
\tag{3.121c}
$$

$$
\begin{aligned}
\bar{\eta}_{1\,\Delta\omega} &= -\pi\tilde{\omega} \int {\rm d}k \int_{-1}^{1} {\rm d}X \int {\rm d}\omega \frac{k^2(1 + X^2)\mathcal{F}_0(\omega^2, k)}{| \gamma_u |^4 |\gamma_\eta|^4} \\
&\quad \times \left\{ 2\omega^4 + \omega^2[\nu(\nu - \eta)k^4 - 2k^2\langle B \rangle^2 X^2] - (\nu + \eta)\eta k^4[\nu\eta k^4 + k^2\langle B \rangle^2 X^2] \right\} \\
&= -\frac{\pi\tilde{\omega}}{(\nu\eta)^{3/2}} \int \frac{{\rm d}k}{k^4} \int_{-1}^{1} {\rm d}X \int {\rm d}\varpi \frac{\mathcal{F}_0(\varpi^2, k)(1 + X^2)}{\mathcal{Z}^2} \\
&\quad \times [2\varpi^4 + \varpi^2(Pm - 1 - 2\mathcal{M}^2 X^2) - (1 + Pm^{-1})(1 + \mathcal{M}^2 X^2)],
\end{aligned}
\tag{3.121d}
$$

$$
\begin{aligned}
\bar{\eta}_{2\,st} &= -\pi\eta\langle B \rangle^2 \int {\rm d}k \int_{-1}^{1} {\rm d}X \int {\rm d}\omega \frac{k^6(13X^4 - 10X^2 + 1)\mathcal{F}_0(\omega^2, k)}{| \gamma_u |^4 |\gamma_\eta|^4} \\
&\quad \times (\omega^2 - \nu\eta k^4 - k^2\langle B \rangle^2 X^2) \\
&= -\frac{\pi}{\nu(\nu\eta)^{3/2}} \int \frac{{\rm d}k}{k^4} \int_{-1}^{1} {\rm d}X \int {\rm d}\varpi \frac{\mathcal{F}_0(\varpi^2, k)(13X^4 - 10X^2 + 1)}{\mathcal{Z}^2} \\
&\quad \times (\varpi^2 - 1 - \mathcal{M}^2 X^2),
\end{aligned}
\tag{3.121e}
$$

and

$$\bar{\eta}_{2\,\Delta\omega} = \frac{3\pi}{2}\frac{\tilde{\omega}\langle B\rangle^2}{(\nu\eta)^{5/2}}\int\frac{\mathrm{d}k}{k^6}\int_{-1}^{1}\mathrm{d}X\int\mathrm{d}\varpi\frac{\mathcal{F}_0(\omega^2,k)}{\mathcal{Z}^2}(13X^4-10X^2+1)$$
$$\times\left\{(2\varpi^2+1+Pm^{-1})-\frac{4(1+Pm^{-1})(Pm+3)\varpi^2(\varpi^2+Pm^{-1})}{3\mathcal{Z}}\right. \tag{3.121f}$$
$$\left.-\frac{4(Pm-Pm^{-1})\varpi^2[2(1+Pm^{-1})+\mathcal{M}^2X^2]}{3\mathcal{Z}}\right\}.$$

with

$$\mathcal{Z} = \varpi^4 + \varpi^2[(Pm+Pm^{-1}) - 2\mathcal{M}^2X^2] + (1+\mathcal{M}^2X^2)^2. \tag{3.122}$$

The induction equation can be cast in the following form

$$\frac{\partial\langle\mathbf{B}\rangle}{\partial t} = \nabla\times(\bar{\alpha}_{\mathrm{ns}}\langle\mathbf{B}\rangle) - \nabla\times(\bar{\eta}_{\mathrm{ns}}\nabla\times\langle\mathbf{B}\rangle), \tag{3.123}$$

where, $\bar{\eta}_{\mathrm{ns}}$ is the full effective magnetic diffusivity, including molecular effects with

$$\bar{\alpha}_{\mathrm{ns}} = \bar{\alpha}_{\mathrm{st}} + \bar{\alpha}_{\mathrm{st}}\Gamma\sin(\tilde{\omega}t) - \bar{\alpha}_{\Delta\omega}\Gamma\cos(\tilde{\omega}t), \tag{3.124a}$$

$$\bar{\eta}_{\mathrm{ns}} = \eta + \bar{\eta}_{\mathrm{st}} + \bar{\eta}_{\mathrm{st}}\Gamma\sin(\tilde{\omega}t) - \bar{\eta}_{\Delta\omega}\Gamma\cos(\tilde{\omega}t). \tag{3.124b}$$

The coefficient $\bar{\eta}_{\mathrm{st}}$ defined in (3.109b) satisfies

$$\bar{\eta}_{\mathrm{st}} = \bar{\eta}_{1\,st} - \bar{\eta}_{2\,st},\quad \text{and}\quad \bar{\eta}_{\Delta\omega} = \bar{\eta}_{1\,\Delta\omega} - \bar{\eta}_{2\,\Delta\omega}. \tag{3.125}$$

The equations (3.124a and b) can be easily rearranged into

$$\bar{\alpha}_{\mathrm{ns}} = \bar{\alpha}_{\mathrm{st}} + \frac{\bar{\alpha}_{\mathrm{st}}\Gamma}{\cos\phi_\alpha}\sin\left(\tilde{\omega}t - \phi_\alpha\right), \tag{3.126a}$$

$$\bar{\eta}_{\mathrm{ns}} = \eta + \bar{\eta}_{\mathrm{st}} + \frac{\bar{\eta}_{\mathrm{st}}\Gamma}{\cos\phi_\eta}\sin\left(\tilde{\omega}t - \phi_\eta\right), \tag{3.126b}$$

where the phases are given by

$$\tan\phi_\alpha = \frac{\bar{\alpha}_{\Delta\omega}}{\bar{\alpha}_{\mathrm{st}}},\qquad \tan\phi_\eta = \frac{\bar{\eta}_{\Delta\omega}}{\bar{\eta}_{\mathrm{st}}}. \tag{3.127}$$

3.3.2.3 An example case: explicit form of the mean EMF in non-stationary, strongly magnetized weak turbulence at low Prandtl number

Let us consider an example case, when the explicit evaluation of the integrals (3.121a–f) is achievable, and on top of (3.115) we also assume a somewhat geophysically relevant limit of strongly magnetized flow with $\nu \ll \eta$, i.e.

$$\forall_k \ \mathcal{M}^2 = \frac{\langle B \rangle^2}{\nu \eta k^2} \gg 1, \qquad Pm = \frac{\nu}{\eta} \ll 1, \qquad Pm\mathcal{M}^2(k) \ll 1, \tag{3.128}$$

where $\mathcal{M}(k)$ is the Hartmann number based on fluctuational length scales and Pm is the magnetic Prandtl number. The last assumption in (3.128),

$$\forall_k \ Pm\mathcal{M}^2(k) \ll 1, \tag{3.129}$$

is only introduced to allow analytical progress and its role is to put an upper bound on the magnetic field strength. Since in the Earth's core the Hartmann number

$$M = \frac{\langle B \rangle L}{\sqrt{\nu \eta}}, \tag{3.130}$$

based on the core depth L is of the order 10^7 and $Pm \approx 5 \times 10^{-7}$ the latter assumption $Pm\mathcal{M}^2(k) \ll 1$ for the Hartmann number $\mathcal{M}(k)$ based on fluctuational wavelengths is not necessarily satisfied for all k. Still, it is reasonable at least in a large part of the short-wavelength spectrum and thus it is put forward in order to simplify the calculations. Note, that since we consider the *strong field* limit, it is not applicable to the linear dynamo regime, i.e. quenching of the electromotive force is significant. Let us take the following simple statistical model (cf e.g. [24, 25], etc)

$$\mathcal{F}_0(\omega^2, k) = \frac{D_0}{k^3}, \qquad \mathcal{F}_1(\omega^2, k) = \frac{D_1}{k^5}, \tag{3.131}$$

with $D_0 = \text{const} > 0$, $|D_1| = \text{const} \leqslant K_e D_0$ (cf [2])[4], which now allows us to calculate the asymptotic forms of the coefficients $\bar{\alpha}_{\text{st}}$, $\bar{\eta}_{\text{st}}$, $\bar{\alpha}_{\Delta\omega}$ and $\bar{\eta}_{\Delta\omega}$ in a straightforward way

---

[4] More precisely, the assumption that force correlations are given by

$$\left\langle \hat{f}_i(\mathbf{k}, \omega)\hat{f}_j(\mathbf{k}', \omega') \right\rangle = \left[\frac{D_0}{k^3} P_{ij}(\mathbf{k}) + \mathrm{i}\frac{D_1}{k^5}\epsilon_{ijk}k_k \right] \delta(\mathbf{k} + \mathbf{k}')\Delta(\omega, \omega'; \tilde{\omega}, \Gamma), \tag{3.132}$$

where the non-stationarity function $\Delta(\omega, \omega'; \tilde{\omega}, \Gamma)$ is defined in (3.86) involves constraints on the parameters D_0, D_1 and Γ. Firstly, calculation of the positive definite quantity $\int_{\delta V_{-\mathbf{q}}} \mathrm{d}^4 q' \left\langle \hat{f}_i(\mathbf{k}, \omega)\hat{f}_i(\mathbf{k}', \omega') \right\rangle = 2D_0/k^3 > 0$, where $\delta V_{-\mathbf{q}}$ denotes the nearest vicinity of $-\mathbf{q}$ in the spectral space (not including $-\omega \pm \tilde{\omega}$), shows that D_0 must be positive. Secondly, the Shwarz inequality (below $\mathrm{d}\mathring{\Omega}$ denotes the solid angle for the $\mathbf{k}$-space)

$$\left| \int \epsilon_{ijk}k_j\hat{f}_k(\mathbf{q})\hat{f}_i(-\mathbf{q})\mathrm{d}\mathring{\Omega} \right|^2 \leqslant \int \left(\epsilon_{ijk}k_j\hat{f}_k(\mathbf{q})\epsilon_{imn}k_m\hat{f}_n(-\mathbf{q}) \right)\mathrm{d}\mathring{\Omega} \int (\hat{f}_i(\mathbf{q})\hat{f}_i(-\mathbf{q}))\mathrm{d}\mathring{\Omega},$$

for the two vectors $\epsilon_{ijk}k_j\hat{f}_k(\mathbf{q})$ and $\hat{f}_i(-\mathbf{q}) = \hat{f}_i^*(\mathbf{q})$ implies $|D_1| \leqslant kD_0$ for all values of k. Finally, calculation of the correlations $\left\langle f_i(\mathbf{x}, t)f_i(\mathbf{x}, t) \right\rangle$ in the real space leads to $|\Gamma| \leqslant 1$, cf (3.112). Summarizing, the following constraints must be imposed on the parameters, which define the forcing in (3.132)

$$D_0 > 0, \qquad |D_1| \leqslant kD_0, \qquad |\Gamma| \leqslant 1. \tag{3.133}$$

Note, however, that $D_1 = -\mathrm{i}k^3\mathbf{k} \cdot \left\langle \hat{\mathbf{f}}(\mathbf{q}) \times \hat{\mathbf{f}}(\mathbf{q})^* \right\rangle /2$ is simply associated with helicity of the forcing and thus can be either positive or negative, imposing the chosen sign of helicity in the turbulent flow. The parameters Γ and $\tilde{\omega}$ can also be assumed either positive or negative. Finally, through (3.79), i.e. $\hat{\mathbf{u}} \approx \hat{\mathbf{f}}/\gamma_u$, the parameters D_0 and D_1 can be associated with the turbulent kinetic energy $\langle \mathbf{u}'^2 \rangle$ and kinetic helicity $\langle \mathbf{u}' \cdot \nabla \times \mathbf{u}' \rangle$, respectively.

$$\bar{\alpha}_{\text{st}} \approx \frac{4\pi^2 D_1}{K_e^2 \langle B \rangle^2}, \qquad \bar{\alpha}_{\Delta\omega} \approx \frac{\pi^3 D_1 \tilde{\omega}}{3\sqrt{Pm}\, K_e^3 \langle B \rangle^3}, \tag{3.134a}$$

$$\bar{\eta}_{1\,st} \approx \frac{\pi^3 D_0}{3\sqrt{\nu\eta}\, K_e^3 \langle B \rangle}, \qquad \bar{\eta}_{1\,\Delta\omega} \approx \frac{\pi^3 D_0 \tilde{\omega}}{20\sqrt{Pm}\, \nu\eta K_e^5 \langle B \rangle}, \tag{3.134b}$$

$$\bar{\eta}_{2\,st} \approx \frac{\pi^3 D_0 \langle B \rangle}{20(\nu\eta)^{3/2} K_e^5}, \tag{3.134c}$$

$$\begin{aligned} &\bar{\eta}_{2\,\Delta\omega} \\ &\approx \frac{3\pi D_0 \tilde{\omega} \langle B \rangle^2}{2(\nu\eta)^{5/2}} \int_{K_e}^{K_\eta} \frac{\mathrm{d}k}{k^9} \int_{-1}^{1} \mathrm{d}X \int \mathrm{d}\varpi \frac{(13X^4 - 10X^2 + 1)(2\varpi^2 + 1 + Pm^{-1})}{\mathcal{Z}^2} \\ &\quad - \frac{2\pi D_0 \langle B \rangle^2}{(\nu\eta)^{7/2}} \eta^2 \tilde{\omega} \int_{K_e}^{K_\eta} \frac{\mathrm{d}k}{k^9} \int_{-1}^{1} \mathrm{d}X \int \mathrm{d}\varpi \frac{(13X^4 - 10X^2 + 1)\varpi^2 \tilde{\mathcal{Z}}}{\mathcal{Z}^3} \\ &\approx \frac{3\pi^3 D_0 \tilde{\omega} \langle B \rangle}{16\sqrt{Pm}(\nu\eta)^2} \int_{K_e}^{K_\eta} \frac{\mathrm{d}k}{k^8} \approx \frac{3\pi^3 D_0 \tilde{\omega} \langle B \rangle}{112\sqrt{Pm}(\nu\eta)^2 K_e^7}, \end{aligned} \tag{3.134d}$$

where

$$\tilde{\mathcal{Z}} = 3\varpi^2 + Pm^{-1} - 2 - \mathcal{M}^2 X^2, \tag{3.135}$$

and we have assumed that the most energetic turbulent eddies are much larger than those on dissipative length scales, $K_e \ll K_\eta$ and by the assumption $Pm \ll 1$ we naturally have $K_\nu \ll K_\eta$; since the calculation of the $\bar{\eta}_{2\,\Delta\omega}$ coefficient was slightly more involving, the intermediate step is shown for clarity.

As mentioned, the purpose of this short subsection is to clearly demonstrate how the turbulent diffusivity and the α coefficient can be calculated in an example case of non-stationary turbulence and in particular we have shown that the significant phase shift in the slow time dependence between the turbulent diffusivity and the α coefficient is a rather general feature of non-stationary turbulence. Under the current assumptions (3.128) the phase shifts (3.127) can be calculated to yield

$$\tan\phi_\alpha = \frac{\bar{\alpha}_{\Delta\omega}}{\bar{\alpha}_{\text{st}}} \approx \frac{\tilde{\omega}\ell_e}{24\sqrt{Pm}\langle B \rangle} = \frac{1}{24}\frac{\tilde{\omega}\ell_e^2}{\nu}\frac{\sqrt{\nu\eta}}{\langle B \rangle \ell_e} = \frac{1}{24}\frac{\tilde{\omega}\ell_e^2}{\nu}\frac{1}{M_{\ell_e}}, \tag{3.136a}$$

$$\tan\phi_\eta = \frac{\bar{\eta}_{\Delta\omega}}{\bar{\eta}_{\text{st}}} \approx \frac{15}{112\pi^2}\frac{\tilde{\omega}\ell_e^2}{\nu}, \tag{3.136b}$$

where

$$M_{\ell_e} = \frac{\langle B \rangle \ell_e}{\sqrt{\nu\eta}} \gg 1, \tag{3.137}$$

is the Hartmann number based on the maximal fluctuational length scale, which in the Earth's core can be expected to be on the order of 10^4 or even 10^5. It follows, that the turbulent diffusion phase ϕ_η is at least a few orders of magnitude larger than the α-effect phase ϕ_α,

$$\phi_\eta \gg \phi_\alpha, \tag{3.138}$$

hence there clearly exists a phase shift between the α-effect and the turbulent magnetic diffusivity. We emphasize, that ϕ_η is proportional to the parameter $\tilde{\omega}$ (frequency shift between distinct waves) e.g. the slowly evolving MAC/magnetic Rossby waves, which is not uniquely determined, and in fact more realistically the non-stationary turbulence should be modelled with a superposition of at least a few (say N) distinct values $\tilde{\omega}_n$, $n = 1, \ldots, N$, each generating a different phase shift $\phi_{\eta\, n}$ between the α-effect and the turbulent diffusivity.

3.3.2.4 Toy model—energy of a force-free mode at $\langle \mathbf{U} \rangle = 0$

The aim of the simple calculation shown in this subsection is to demonstrate how a phase shift between the field-amplifying α-effect and the resistive decay can lead to 'magnetic field excursions', that is, short periods of magnetic energy decay separating longer periods characterized by relatively stable magnetic energy. We will demonstrate an explicit numerical solution obtained for a force-free mode defined by

$$\nabla \times \langle \mathbf{B} \rangle = \kappa \langle \mathbf{B} \rangle, \tag{3.139}$$

where κ is a scalar constant; this ensures vanishing of the Lorentz force $\nabla \times \langle \mathbf{B} \rangle \times \langle \mathbf{B} \rangle = 0$. Such force-free states are known to exist and have been intensively investigated, e.g. in seminal works of Chandrasekhar (1956) [26] and Chandrasekhar and Kendall (1957) [27] (cf also more recent works [28] and [29]). On defining the magnetic energy

$$E_m = \frac{1}{2}\langle B \rangle^2, \tag{3.140}$$

the induction equation (3.123) in the absence of large-scale flows, $\langle \mathbf{U} \rangle = 0$, yields for the energy

$$\frac{\partial E_m}{\partial t} = 2\bar{\alpha}\kappa(\mathbf{x})E_m - 2\bar{\eta}\kappa(\mathbf{x})^2 E_m. \tag{3.141}$$

Based on the previous analysis we may propose the following

$$\begin{aligned} \bar{\alpha} &= \frac{\bar{\alpha}_0}{1 + \bar{A}E_m^2}\left[1 + \Gamma\cos\left(\tilde{\omega}t - \phi_\alpha\right)\right], \\ \bar{\eta} &= \eta + \frac{\bar{\eta}_0}{1 + \bar{A}E_m^2}\left[1 + \Gamma\cos\left(\tilde{\omega}t - \phi_\eta\right)\right], \end{aligned} \tag{3.142}$$

where instead of the magnetic field dependence obtained in (3.134–d) valid only in the strong field limit we added a standard quenching factor, which models action of the Lorentz force in the simplest way (cf [13, 30]), with $\bar{A} = \text{const}$. Taking

$$\phi_\alpha = 0, \qquad \phi_\eta = \pi, \tag{3.143}$$

gives a well suited case study, since it implies that the maximal enhancement of the field takes place when the effect of the resistive decay is the weakest and vice versa—the strongest resistive decay of the mean field is associated with the weakest amplification by the α-effect; the latter situation leads to possible excursions (field

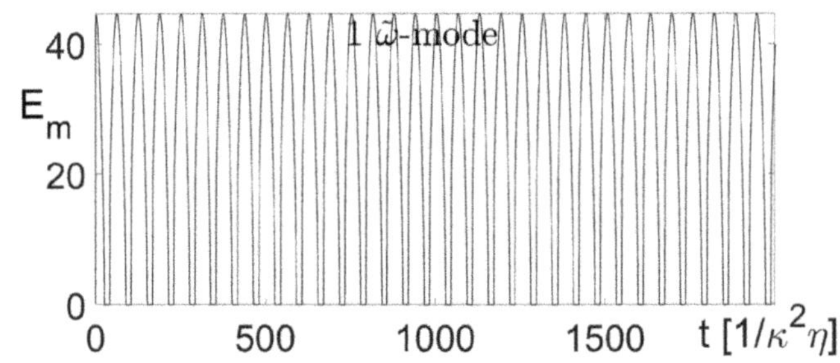

Figure 3.1. Evolution of the magnetic energy in non-stationary turbulence according to equation (3.144), when non-stationary force correlations $\langle f_i(\mathbf{x}, t) f_i(\mathbf{x}', t) \rangle \sim 1 + \sin(\tilde{\omega} t)$ are defined by a single-mode time dependence with the frequency $\tilde{\omega} = 0.0001$; the remaining parameters are $\eta = 0.001$, $\bar{\eta}_0 = 0.1$, $\bar{\alpha}_0 = 1$, $\bar{A} = 1$, $\kappa = 1$. The plot is shown in diffusive time units $1/\kappa^2\eta = 10^3$.

decay), whereas the former one to 'stable' periods with strong magnetic field. In such a case we get

$$\frac{\partial E_m}{\partial t} = 2\bar{\alpha}_0 \kappa (1 + \cos \tilde{\omega} t) \frac{E_m}{1 + \bar{A} E_m^2}, \\ - 2\kappa^2 \left[\eta \left(1 + \bar{A} E_m^2\right) + \bar{\eta}_0 - \bar{\eta}_0 \cos(\tilde{\omega} t) \right] \frac{E_m}{1 + \bar{A} E_m^2}, \tag{3.144}$$

where we have inserted $\Gamma = 1$. A selected solution of the latter equation is shown in figure 3.1, where indeed regular short-lived excursions are manifested in the evolution of the magnetic energy, i.e. short periods with suppressed magnetic field. More generally, accounting for the possibility of a more complex time dependence of the force correlations with more than one slow $\tilde{\omega}$-mode, the magnetic energy evolution equation can be rewritten in the form

$$\frac{\partial E_m}{\partial t} = 2\bar{\alpha}_0 \kappa \left[1 + \sum_n \Gamma_n \cos(\tilde{\omega}_n t + \psi_n) \right] \frac{E_m}{1 + \bar{A} E_m^2} \\ - 2\kappa^2 \left[\eta \left(1 + \bar{A} E_m^2\right) + \bar{\eta}_0 + \bar{\eta}_0 \sum_n \Gamma_n \cos\left(\tilde{\omega}_n t + \phi_n\right) \right] \frac{E_m}{1 + \bar{A} E_m^2}. \tag{3.145}$$

Selected solutions are shown in figure 3.2 for the cases of four and eight $\tilde{\omega}$-modes. It is evident, that longer periods of relative stability of the magnetic energy are separated by much shorter excursions and this effect is more pronounced as the number of the $\tilde{\omega}$-modes increases.

3.3.3 Discussion

The studied problem of weak turbulence stirred by non-stationary forcing is obviously a great simplification. It relies on derivations obtained under restrictive assumptions of statistic homogeneity and isotropy and linearized evolution of the turbulent pulsations. In addition, the toy model considered in the end was focused

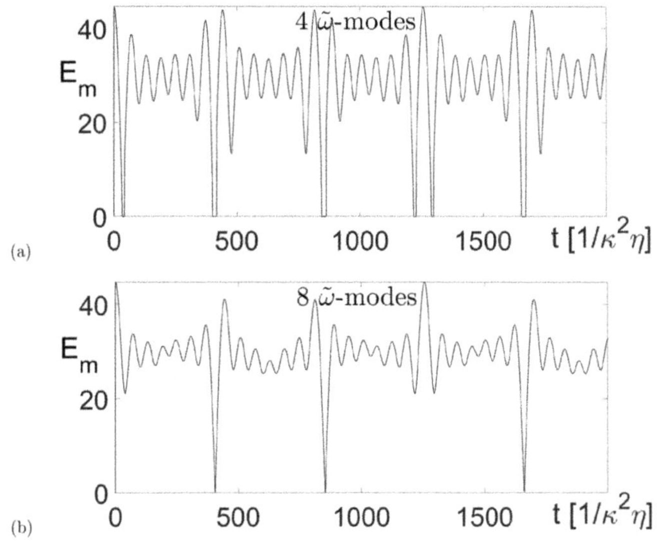

Figure 3.2. Evolution of the magnetic energy in non-stationary turbulence according to equation (3.145) with $\psi_n = 0$, $\phi_n = \pi$, when non-stationary force correlations $\langle f_i(\mathbf{x}, t) f_i(\mathbf{x}', t) \rangle$ are defined by a four-mode time dependence with the frequencies and amplitudes ($\tilde{\omega}_1 = 0.0001$, $\Gamma_1 = 0.4$), ($\tilde{\omega}_2 = 0.000\,085$, $\Gamma_2 = 0.3$), ($\tilde{\omega}_3 = 0.000\,07$, $\Gamma_3 = 0.2$), ($\tilde{\omega}_4 = 0.000\,055$, $\Gamma_4 = 0.1$) (a) and an eight-mode time dependence with ($\tilde{\omega}_1 = 0.0001$, $\Gamma_1 = 0.2$), ($\tilde{\omega}_2 = 0.000\,085$, $\Gamma_2 = 0.2$), ($\tilde{\omega}_3 = 0.000\,07$, $\Gamma_3 = 0.15$), ($\tilde{\omega}_4 = 0.000\,055$, $\Gamma_4 = 0.15$), ($\tilde{\omega}_5 = 0.000\,04$, $\Gamma_5 = 0.1$), ($\tilde{\omega}_6 = 0.000\,025$, $\Gamma_6 = 0.1$), ($\tilde{\omega}_7 = 0.000\,015$, $\Gamma_7 = 0.05$), and ($\tilde{\omega}_8 = 0.000\,005$, $\Gamma_8 = 0.05$) (b); the remaining parameters are $\eta = 0.001$, $\bar{\eta}_0 = 0.1$, $\bar{\alpha}_0 = 1$, $\bar{A} = 1$, $\kappa = 1$. The plot is shown in diffusive time units $1/\kappa^2\eta = 10^3$.

on the force-free magnetic modes only. However, this model is only used to visualize the studied effect of non-stationarity of the turbulent coefficients $\bar{\alpha}$ and $\bar{\eta}$, which is generic and its existence is independent of the simplifying assumptions. *Their temporal evolution was shown to be out of phase, implying existence of periods of enhanced and suppressed turbulent dynamo process, which may correlate or not with periods of enhanced/suppressed diffusion. In particular, enhancement of the α-effect may coincide with suppression of diffusion (stable field) and vice versa, enhancement of diffusion may coincide with suppression of the dynamo effect (field decay, excursion). It would be instructive to study such temporal characteristics of the turbulent diffusivity and the α-effect in numerical simulations, in particular the correlations between the value of the ratio $\bar{\eta}/\bar{\alpha}$ and the occurrences of excursions.*

As shown, in a non-stationary turbulence an enhancement of diffusion may be correlated with simultaneous suppression of the α-effect, in which case it is likely to result in a sharp drop of the large-scale magnetic field intensity, i.e. the excursion. With such a picture the geomagnetic excursions are manifestations of the chaotic core turbulence and their occurrences are also bound to be chaotic, and no characteristic, significant alterations in the core large-scale flow are expected during excursions, in accordance with results of numerical simulations of [31–33]; a schematic picture of the internal structure of the Earth depicting the location depth

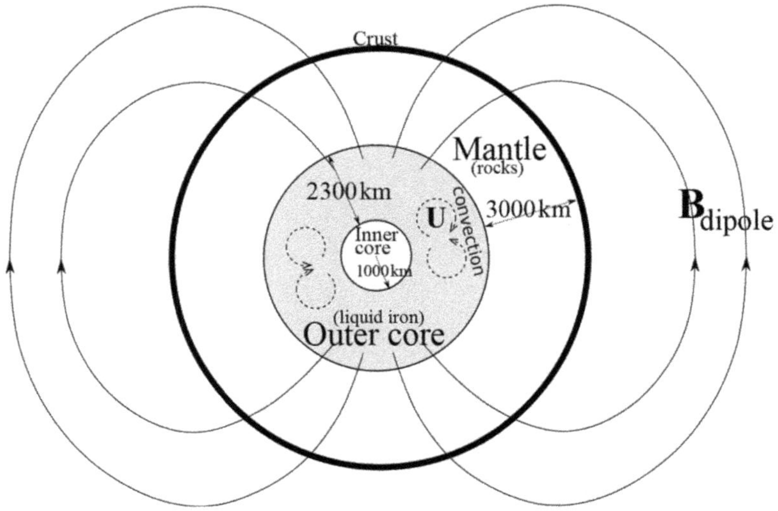

Figure 3.3. An illustrative diagram of the internal structure of the Earth. The rocky mantle lies above the liquid iron core, where convective flows driven by primordial heat flux accumulated at planet's formation generate the geomagnetic field via the dynamo effect. Although inside the core the magnetic field has a very complex, tangled structure, at the Earth's surface, it is predominantly dipolar. The direction of the magnetic lines of force corresponds to the current epoch.

and thickness of the liquid iron core where the geomagnetic field is generated is shown in figure 3.3. Perhaps with the development of the available computational abilities and power, in future it may become possible to numerically study the Earth-like parameter regime and construct reliable fits to the geomagnetic data in order to provide estimates of the turbulent magnetic diffusivity and the α-effect, which could then be monitored since a drop in the value of the ratio $\bar{\alpha}/\bar{\eta}$ for the outer core could indicate higher possibility for a geomagnetic excursion.

The excursions and in particular polarity reversals have been observed in numerical simulations. It was reported that increasing the vigour of convection tends to increase the frequency of polarity reversals (cf [31, 32]). In simulations of [33] the first stage of a polarity reversal is characterized by intensification of turbulence in some region of the core where the vigorous flow twists and bends the field lines to locally reverse the magnetic field direction. Since the magnitude of the turbulent flow in that region is particularly vigorous, the local magnetic Reynolds number is high, thus favouring conditions for amplification of the reversed field, which spreads into surrounding regions until the polarity in the core becomes mixed. This implies sharp decay of the dipole moment. Consequently, we may expect that the magnetic excursions and polarity reversals are manifestations of a chaotic turbulent behaviour of the liquid core and no significant large-scale phenomena seem to take place in the flow during excursions.

It needs to be stressed, that in the presented analysis the large-scale EMF is calculated with the use of fluctuational equations, but then we concentrate on the evolution of the large-scale field leaving the dynamics of the small-scale component of the magnetic field unexplored. It follows, that temporal enhancement of turbulent magnetic diffusivity does not need to suppress the small-scale dynamo, which in reality is actually rather expected to be vividly operating on much shorter timescales, coupling with the large-scale process. Furthermore, as stated above, the problem is greatly simplified from the start by assumption of isotropy and neglection of the strongly anisotropic effect of the background rotation and density stratification. At first rough estimate, the anisotropy leads to separation between horizontal and vertical turbulent transport coefficients, i.e. $\bar{\alpha}_{ij} = \bar{\alpha}_h\delta_{ij} + (\bar{\alpha}_v - \bar{\alpha}_h)\delta_{i3}\delta_{j3}$ and $\bar{\eta}_{ij} = \bar{\eta}_h\delta_{ij} + (\bar{\eta}_v - \bar{\eta}_h)\delta_{i3}\delta_{j3}$ (say rotation is along the z-axis). In fact it was shown in [34] with the use of the renormalization group method, that in the limit of rapid rotation with $Pm \ll 1$ strong turbulence tends to enhance the vertical $\bar{\alpha}_v$ coefficient with respect to the horizontal one $\bar{\alpha}_h$ and makes them comparable, in contrast to the weak turbulence limit when $\bar{\alpha}_h \gg \bar{\alpha}_v$[5]. It is also known, that in the weak seed field limit (linear stage of growth of the mean magnetic field) the mean induction equation with anisotropic $\bar{\boldsymbol{\alpha}}$ and $\bar{\boldsymbol{\eta}}$ tensors separates the evolution of even and odd modes, thus dipolar parity evolves independently of the quadrupolar one.

Moreover, one should bear in mind that the reality is even more complex—not only non-stationary but also inhomogeneous, leading to spatially dependent turbulent coefficients, which further differentiates the evolution of different spatial modes. Summarizing, the small-scale dynamo is by no means excluded during the periods of enhanced turbulent diffusion and the simple isotropic, homogeneous model obviously does not grasp the complexity of the strongly anisotropic turbulence and the mean-field evolution in many natural systems such as e.g. the Earth's core; and it is the anisotropy and non-homogeneity (rotation, stratification, buoyancy, inhomogeneous wave interactions, etc) that lead to significant differences in the evolution of the mean dipolar and quadrupolar fields. Although theoretical models including non-stationarity, inhomogeneity and anisotropy may be too cumbersome to investigate, it could be of interest to utilize DNS (*D*irect *N*umerical *S*imulations) to study the long time evolution of turbulent coefficients in spherical, rapidly rotating shells and its dynamical connection to magnetic excursions.

On a final note, we comment on the relative signs of the standard diffusive contribution to the α-effect and the non-equilibrium one in order to understand when the two largely independent effects cooperate to increase the efficiency of the dynamo process and when they compete and weaken the dynamo. In section 3.2 on forced beating waves we have shown that the relative sign of the two effects depends on the phase shift between the two components of the forcing which oscillate at

[5] With the use of narrow spherical gap limit and under neglection of gravity and density stratification, Mizerski [34] showed, that the growth rate of the mean magnetic field is latitude dependent and the results suggest that the mean field should be amplified near the poles up to mid-latitudes, but damped near the equator. Although the narrow gap limit is naturally a great simplification, it is somewhat justified by the dynamical features of the Earth's outer core. The so-called *S*tratified *O*cean at the top of the *C*ore (SOC), i.e. a thin strongly stratified layer at the top of the Earth's liquid core is regarded as dynamically crucial in the geodynamo theory (cf [35, 36]).

distinct but close frequencies. A phase shift of $\pi/2$ with $\hat{\mathbf{f}}_1 = \mathrm{i}\hat{\mathbf{f}}_2$ leads to the same signs of the diffusive and non-equilibrium contributions to the α-effect, but of course an opposite shift $\hat{\mathbf{f}}_1 = -\mathrm{i}\hat{\mathbf{f}}_2$ would lead to competition between them. Hence the relative sign depends on the parameters of the forcing, that is the input parameters which we control. Similarly, when turbulence is stirred by a Gaussian non-stationary forcing of the form (3.132) the results (3.124) and (3.134) clearly demonstrate cooperation when $\Gamma\tilde{\omega}<0$ and competition when $\Gamma\tilde{\omega}>0$ in the initial stage of evolution $t \ll \tilde{\omega}^{-1}$, i.e. the joint action of the two effects—the diffusive $\bar{\alpha}_{\mathrm{st}}$ and the non-equilibrium $\bar{\alpha}_{\Delta\omega}$—is controlled by the parameters of the stirring force. It is important to realize, that when the two effects compete, one of them needs to dominate over the other in order to get an α-type dynamo and in the case when the standard diffusive dynamo is very efficient $|\bar{\alpha}_{\mathrm{st}}| \gg |\bar{\alpha}_{\Delta\omega}|$ and $\bar{\alpha}_{\mathrm{st}}\bar{\alpha}_{\Delta\omega} < 0$, not very strong non-equilibrium effects would only lead to attenuation of the α-effect.

3.3.4 Nonlinear effects of the Lorentz force

Finally, it is instructive to comment on the non-obvious effect of the Lorentz force, visible in the expressions for the turbulent magnetic diffusivity obtained in section 3.3.2.3. The total turbulent diffusivity coefficient $\bar{\eta}_{\mathrm{ns}}$ defined in (3.124b), (3.125) and (3.121c–f) takes the form

$$\bar{\eta}_{\mathrm{ns}} = \eta + \bar{\eta}_{1\,st} - \bar{\eta}_{2\,st} + (\bar{\eta}_{1\,st} - \bar{\eta}_{2\,st})\Gamma\sin(\tilde{\omega}t) + (\bar{\eta}_{2\,\Delta\omega} - \bar{\eta}_{1\,\Delta\omega})\Gamma\cos(\tilde{\omega}t). \quad (3.147)$$

Clearly, with the results of section 3.3.2.3 we can see, that in the limit of strong magnetic field and low magnetic Prandtl number

$$M_{\ell_e} = \frac{\langle B\rangle\ell_e}{\sqrt{\nu\eta}} \gg 1, \qquad PmM_{\ell_e} \ll 1, \quad (3.148)$$

there exists a negative contribution to the turbulent magnetic diffusivity

$$-\bar{\eta}_{2\,st} \approx -\frac{D_0\ell^4}{640\pi^2\eta^2}Pm^{-1}M_{\ell_e} < 0, \qquad \bar{\eta}_{2\,st} \gg \bar{\eta}_{1\,st}. \quad (3.149)$$

This is an interesting observation as this coefficient describes nonlinear diffusion, and because it depends on the magnetic field $\langle B\rangle$, thus it must vanish in the absence of the Lorentz force. Of course in the considered limit (3.148) the Lorentz force must be large. It follows, that in this case the effect of the Lorentz force is non-standard, since it creates a significant decrease of the magnetic diffusivity, which tends to delay saturation of the magnetic energy. However, it is clear, that the assumptions (3.148) put an upper bound on the magnetic field strength, $M_{\ell_e} \ll Pm^{-1}$, therefore, in the case when the magnetic energy is enhanced by the dynamo effect above this constraint, the derived structure of the EMF, i.e. the α-effect and the turbulent diffusivity is lost. Moreover, we must bear in mind that the results in this chapter were obtained within the limit of weak turbulence, that is, under linearized evolution of the turbulent fluctuations. It follows that all the turbulent contributions to the effective magnetic diffusivity, i.e. $\bar{\eta}_{1\,st}$, $\bar{\eta}_{2\,st}$, $\bar{\eta}_{1\,\Delta\omega}$ and $\bar{\eta}_{2\,\Delta\omega}$ may be weaker than the molecular diffusivity η.

The sign of the non-stationary contribution

$$(\bar{\eta}_{2\,\Delta\omega} - \bar{\eta}_{1\,\Delta\omega})\Gamma\cos(\tilde{\omega}t) \approx \bar{\eta}_{2\,\Delta\omega}\Gamma \approx \frac{3D_0\ell_e^4}{14\,336\pi^4\eta^2}\frac{\tilde{\omega}\ell_e^2}{\eta}Pm^{-2}M_{\ell_e}\Gamma, \qquad (3.150)$$

in the considered limit (3.148) is positive within the initial stage of evolution $t \ll \tilde{\omega}^{-1}$.

3.4 Non-equilibrium dynamo effects in low-resistivity plasma of interstellar medium (ISM) and active galactic nuclei (AGN)

In hot plasma, the density ρ is extremely low and thus the kinematic viscosity $\nu = \mu/\rho$ is high. This is because the dynamic viscosity μ is almost independent of density in the limit $\rho \to 0$, as already recognized by Maxwell (1867) [37]. It follows that $\eta \ll \nu$ is satisfied, or that the magnetic Prandtl number $Pm = \nu/\eta \gg 1$ is large. The large-scale hydromagnetic dynamo mechanism which relies on the existence of a resistive phase shift between the kinetic and magnetic components of waves is therefore unlikely to effectively operate in hot, low-density plasma, where the electrical resistivity is extremely low. In this section it is shown that non-stationarity, a common factor in plasma turbulence, e.g. in the interstellar medium caused by random supernova explosions, allows inducing a strong large-scale EMF by the plasma flow and significantly enhances the dynamo effect. We adopt here the same approach as in the previous section, i.e. the incompressible MHD turbulence is stirred with a random, Gaussian, and non-stationary forcing defined by (3.132). The model considered here is naturally a great simplification with respect to the real flows, which feature differential rotation, density stratification, self-gravity, and so on, however, it allows us to obtain analytic results. The EMF is calculated analytically as in section 3.3.2.3 in the limit of strong magnetic fields, however, contrary to that section, here we assume large magnetic Prandtl number $Pm \gg 1$ (low magnetic diffusion in comparison with plasma viscosity). It is demonstrated that the EMF induced by the non-equilibrium effects can be typically expected to dominate the dynamo mechanism (over the mechanism based on diffusive phase shift) by several orders of magnitude in two example cases of the interstellar medium (ISM) in the Milky Way Galaxy and hot accretion disks such as those of AGN. Despite the great simplifications of the model, the result is generic in the sense that small-scale turbulence, which exhibits chirality and non-stationarity, induces an efficient large-scale dynamo process.

The stirring force $\mathbf{f}$ is assumed homogeneous and isotropic but chiral (helical) and Gaussian with zero mean, $\langle \mathbf{f} \rangle = 0$ and is fully defined by the following correlation function

$$\left\langle \hat{f}_i(\mathbf{k}, \omega)\hat{f}_j(\mathbf{k}', \omega') \right\rangle = \left[\frac{D_0}{k^3}P_{ij}(\mathbf{k}) + \mathrm{i}\frac{D_1}{k^5}\epsilon_{ijk}k_k \right]\delta(\mathbf{k} + \mathbf{k}')\Delta(\omega, \omega'; \tilde{\omega}), \qquad (3.151)$$

where the function $\Delta(\omega, \omega'; \tilde{\omega})$ describes its statistical temporal properties, and is given in (3.86), chosen in such a way that in the non-stationary case, the correlations in real space have a simple sinusoidal time dependence $\langle f_i(\mathbf{x}, t) f_i(\mathbf{x}', t)\rangle \sim \sin(\tilde{\omega} t)$; full description of the expression (3.151) is provided in footnote 4. With such a definition the non-stationarity parameter $\tilde{\omega}$ is proportional to the turbulent energy production rate. The non-helical part of the correlation function, $D_0 P_{ij}(\mathbf{k})/k^3$ is inversely proportional to the third power of the wave number, whereas the helical one, $\mathrm{i}D_1\epsilon_{ijk}k_k/k^5$, scales like k^{-4}; as shown by [24, 25], this particular choice of scalings corresponds to the Kolmogorov-type turbulence when $\Delta(\omega, \omega'; \tilde{\omega}) = \delta(\omega + \omega')$, that is, it reproduces the Kolmogorov-type scaling laws for the spectral kinetic energy $\langle \hat{u}^2(\mathbf{k}, \omega)\rangle \sim k^{-5/3}$ and spectral helicity $\langle \hat{\mathbf{u}}(\mathbf{k}, \omega) \cdot \hat{\mathbf{w}}(\mathbf{k}, \omega)\rangle \sim k^{-5/3}$ for a stationary case in the absence of the magnetic field.

We consider the situation in which the gradients of mean fields are negligibly small in the equations for fluctuations, i.e. the mean fields are effectively spatially homogeneous on scales $\leqslant \ell_e$. For clarity, we neglect all effects associated with the mean flow $\langle \mathbf{U}\rangle$ in order to concentrate solely on the influence of the non-stationarity of turbulence on the α-effect. Thus the equations for the Fourier transforms of the turbulent fluctuations $\hat{\mathbf{u}}$ and $\hat{\mathbf{b}}$ can be written in the form, cf (3.79 and b) with $\mathbf{G} = 0$ and $\boldsymbol{\Gamma} = 0$,

$$\hat{\mathbf{u}}(\mathbf{q}) = \frac{1}{\gamma_u}\hat{\mathbf{f}}(\mathbf{q}), \qquad \hat{\mathbf{b}}(\mathbf{q}) = \mathrm{i}\frac{\mathbf{k} \cdot \langle \mathbf{B}\rangle}{\gamma_\eta}\hat{\mathbf{u}}(\mathbf{q}) = \mathrm{i}\frac{\mathbf{k} \cdot \langle \mathbf{B}\rangle}{\gamma_u\gamma_\eta}\hat{\mathbf{f}}(\mathbf{q}), \tag{3.152}$$

where $\gamma_\nu = -\mathrm{i}\omega + \nu k^2$, $\gamma_\eta = -\mathrm{i}\omega + \eta k^2$, $\gamma_u = \gamma_\nu + (\mathbf{k} \cdot \langle \mathbf{B}\rangle)^2/\gamma_\eta$ and $\mathbf{q} = (\omega, \mathbf{k})$. The neglection of the mean field gradients requires a comment. Although such an approach is common in the mean-field dynamo theory, it eliminates shearing instabilities, since the effect of the mean velocity shear on the dynamics of fluctuations is removed. Hence turbulence is driven solely by the forcing. In addition, the shear-current effect ([38–41]), the so-called cross-helicity dynamo ([42]), and the effects of turbulent diffusion are also excluded by neglection of the mean field gradients, see equation (2.26) and the discussion below. Especially, the latter can be important, as turbulent diffusion could account for an enhancement in the shift between the kinetic and magnetic components of waves and thus induction of the large-scale magnetic field via the standard, well-known diffusive mechanism. Nevertheless, we concentrate here solely on the α-effect, since it is the lowest order effect in terms of the weak derivatives of the large-scale field, and in this sense may be expected to be dominant in some situations (cf [2]). The FOSA assumption, i.e. the assumption of weak turbulence is, in fact, very limiting, since the Reynolds numbers for ISM and AGN are very high; for example, in the case of ISM in the Milky Way Galaxy $Re = 10^8$, $Rm = RePm \approx 10^{22}$ (cf [13, 43]). In such a case the turbulence is necessarily strong, that is, the evolution of fluctuations is bound to be strongly nonlinear. Nevertheless, the assumption of weak turbulence allows one to obtain analytic results in the limit of strong magnetic fields and large Prandtl numbers,

$$Pm = \frac{\nu}{\eta} \gg \mathcal{M}^2(k) = \frac{\langle B\rangle^2}{\nu\eta k^2} \gg 1, \qquad K_e \leqslant k \leqslant K_\nu, K_\eta, \tag{3.153}$$

and thus for a clear estimation of the relative strength of the dynamo effect generated by non-stationarity. Analytic results are rare in dynamo theory, this is why it seems instructive to utilize the simple framework of weak turbulence, despite its obvious limitation.

The large-scale EMF is now expressed by the first term in (3.100) in the following way

$$\epsilon_{ljk} \int d^4q e^{i(\mathbf{k}\cdot\mathbf{x}-\omega t)} \int d^4q' e^{i(\mathbf{k}'\cdot\mathbf{x}-\omega' t)} \left\langle \hat{u}_j(\omega, \mathbf{k})\hat{b}_k(\omega', \mathbf{k}')\right\rangle = \mathcal{I}_1(\nu, \eta, \langle B\rangle^2)\langle B\rangle_l, \tag{3.154a}$$

$$\mathcal{I}_1(\nu, \eta, \langle B\rangle^2) = 4\pi D_1 \int dk \int_{-1}^{1} dX \int d\omega \int d\omega' e^{-i(\omega+\omega')t} \frac{X^2}{k\gamma_u\gamma_u'\gamma_\eta'} \Delta(\omega, \omega'; \tilde{\omega}). \tag{3.154b}$$

By the use of equation (3.120) at the initial stage of the dynamo process defined by

$$t \ll 1/\tilde{\omega}, \qquad \tilde{\omega} \ll \frac{\eta}{\ell_e^2}, \tag{3.155}$$

one obtains

$$\mathcal{I}_1(\nu, \eta, \langle B\rangle^2) = \bar{\alpha}_{st} - \Gamma\bar{\alpha}_{\Delta\omega}, \tag{3.156}$$

where

$$\bar{\alpha}_{st} = \frac{4\pi D_1}{\nu\sqrt{\nu\eta}} \int \frac{dk}{k^5} \int_{-1}^{1} dX \int_{-\infty}^{\infty} d\varpi \frac{X^2}{\mathcal{Z}}, \tag{3.157a}$$

$$\bar{\alpha}_{\Delta\omega} = -\frac{4\pi D_1\tilde{\omega}}{(\nu\eta)^{3/2}} \int \frac{dk}{k^7} \int_{-1}^{1} dX \int d\varpi \frac{X^2}{\mathcal{Z}^2}[2\varpi^4 + \varpi^2(Pm - 1 - 2\mathcal{M}^2X^2) - (1 + Pm^{-1})(1 + \mathcal{M}^2X^2)]. \tag{3.157b}$$

In the asymptotic limit (3.153) one obtains

$$\begin{aligned}\bar{\alpha}_{st} &\approx \frac{4\pi^2 D_1}{\nu^2} \int_{K_e}^{K_\nu} \frac{dk}{k^5} \int_{-1}^{1} \frac{X^2 dX}{(1 + \mathcal{M}^2X^2)} \\ &\approx \frac{8\pi^2 D_1}{Pm\langle B\rangle^2} \int_{K_e}^{K_\nu} \frac{dk}{k^3} \approx \frac{4\pi^2 D_1}{Pm\langle B\rangle^2 K_e^2};\end{aligned} \tag{3.158a}$$

$$\begin{aligned}\bar{\alpha}_{\Delta\omega} &\approx -\frac{2\pi^2 D_1\tilde{\omega}}{\nu^2\eta} \int_{K_e}^{K_\nu} \frac{dk}{k^7} \int_{-1}^{1} dX \frac{\mathcal{M}^4X^2}{(2 + \mathcal{M}^2X^2)(1 + \mathcal{M}^2X^2)^2} \\ &\approx -\frac{(3 - 2\sqrt{2})\pi^3 D_1\tilde{\omega}}{\nu^2\eta} \int_{K_e}^{K_\nu} \frac{\mathcal{M} dk}{k^7} \approx -\frac{(3 - 2\sqrt{2})\pi^3 D_1\tilde{\omega}\langle B\rangle}{7\nu(\nu\eta)^{3/2}K_e^7},\end{aligned} \tag{3.158b}$$

where we have also used $K_e \ll K_\nu \ll K_\eta$ (since $Pm \gg 1$). The assumption in equation (3.155) on the one hand ensures that the evolution of the mean EMF is very slow, much slower than that of the turbulent fluctuations, but on the other hand, it implies weak non-stationarity. Nevertheless, even in this restrictive limit (the time-scale of magnetic diffusion is *extremely* long in the case of ISM and AGN), the α-effect induced by non-stationarity in high-*Pm*, strongly magnetized plasma is likely to be significantly stronger than the stationary, diffusive one since the ratio

$$\left|\frac{\bar{\alpha}_{\Delta\omega}}{\bar{\alpha}_{\rm st}}\right| \approx \frac{3-2\sqrt{2}}{896\pi^4}\frac{\tilde{\omega}\ell_e^2}{\eta}M^3, \qquad \text{where} \qquad M = \frac{\langle B\rangle \ell_e}{\sqrt{\nu\eta}}, \tag{3.159}$$

exceeds unity when

$$\frac{\tilde{\omega}\ell_e^2}{\eta} > \frac{896\pi^4}{3-2\sqrt{2}}M^{-3} \approx \frac{5\times 10^5}{M^3}. \tag{3.160}$$

The latter is, in fact, a rather mild restriction, since by assumption, $M \gg 1$ and even $10^6 M^{-3} < 1$ is typically satisfied in hot, low-density plasma. For example, in the case of the ISM of the Milky Way Galaxy, with the aid of [43] and [13], we can take the following characteristic values:

$$\nu_{\rm ISM} \approx 10^{17}\,\frac{m^2}{s}, \qquad \eta_{\rm ISM} \approx 10^3\,\frac{m^2}{s}, \qquad \langle B\rangle_{\rm ISM} \approx 0.5\,nT, \tag{3.161a}$$

$$\ell_{e\rm ISM} \approx 10^{18}\,m, \qquad \rho_{\rm ISM} \approx 10^{-21}\,\frac{kg}{m^3}, \tag{3.161b}$$

which lead to $Pm_{\rm ISM} \approx 10^{14} \gg 1$, $M_{\rm ISM} \approx 10^{12}$, and hence $5\times 10^5 M^{-3} \approx 5\times 10^{-31}$ for the ISM; $\ell_{\rm ISM}$ was taken as the supernova scale at which the turbulence is stirred (cf also [44]). A second example involves AGN, that is, hot accretion disks around the supermassive central black holes in spiral galaxies, for which we take the following characteristic values (cf [13, 45]):

$$\nu_{\rm AGN} \approx 10^{3}\,\frac{m^2}{s}, \qquad \eta_{\rm AGN} \approx 10^{-1}\,\frac{m^2}{s}, \qquad \langle B\rangle_{\rm AGN} \approx 5\,\mu T, \tag{3.162a}$$

$$\ell_{e\rm AGN} \approx 10^{7}\,m, \qquad \rho_{\rm AGN} \approx 10^{-2}\,\frac{kg}{m^3}. \tag{3.162b}$$

This leads to $Pm_{\rm AGN} \approx 10^4 \gg 1$, $M_{\rm AGN} \approx 10^4$, and hence $5\times 10^5 M^{-3} \approx 5\times 10^{-7}$ for the AGN.

In both example cases, that is, for the ISM and for the AGN, the particular assumption $Pm \gg \mathcal{M}^2$ made in order to facilitate and explicitly calculate the Fourier integrals is not strictly satisfied for all values of the wave number k. Nevertheless, the result highlights the importance of non-stationarity of turbulent magnetized plasma, since even weak non-stationarity leads to a significant enhancement of the α-effect. In particular, in the case of ISM turbulence it is driven by the supernovae ejections and thus the supernovae ejection rate, or more precisely the difference between the ejection rates for type I and type II supernovae, could be adopted as a measure of non-stationarity. From data provided by [46] we can estimate it at $\tilde{\omega}\approx 10^{-10}\ /s$, which by (3.159), (3.160) and (3.161 and b) allows us to argue that the non-equilibrium dynamo effect $\bar{\alpha}_{\Delta\omega}$ strongly dominates the diffusive dynamo effect $\bar{\alpha}_{\rm st}$.

The most important disadvantage of our theory is naturally the simplification of the physical model, involving the assumptions of isotropy and homogeneity of the turbulent flow and the FOSA approximation, which are necessary in order to provide explicit analytic results. Real turbulence in natural hot, low-density plasma is strong (very high Reynolds numbers), anisotropic and inhomogeneous due to the action of differential rotation, gravity, density stratification, etc. Nevertheless, despite the great simplification, the analysis presented in this section clearly shows the importance of non-equilibrium effects in the high-Pm turbulence and the large-scale dynamo process, which could easily dominate the standard α-effect based on creation of a phase shift between the velocity and magnetic turbulent fluctuations by the very weak magnetic diffusion. On a final note, we observe that in this case the non-equilibrium effect $\bar{\alpha}_{\Delta\omega}$ and the diffusive one $\bar{\alpha}_{\rm st}$ cooperate (have the same signs) when $\Gamma\tilde{\omega}>0$, cf (3.156), (3.158) and (3.4).

References

[1] Lehnert B 1954 Magnetohydrodynamic waves under the action of the Coriolis force *Astrophys. J.* **119** 647–54

[2] Moffatt H K and Dormy E 2019 *Self-exciting Fluid Dynamos* (Cambridge: Cambridge University Press)

[3] Tobias S M, Cattaneo F and Boldyrev S 2013 MHD dynamos and turbulence *Ten Chapters in Turbulence* ed P A Davidson, Y Kaneda and K R Sreenivasan (Cambridge: Cambridge University Press) pp 351–404

[4] Zakharov V E, L'vov V S and Falkovich G 1992 *Kolmogorov Spectra of Turbulence* (Berlin: Springer)

[5] Newell A C and Rumpf B 2011 Wave turbulence *Ann. Rev. Fluid Mech.* **43** 59–78

[6] Nazarenko S 2011 *Wave Turbulence* (Berlin: Springer)

[7] Monsalve E, Brunet M, Gallet B and Cortet P-P 2020 Quantitative experimental observation of weak inertial-wave turbulence *Phys. Rev. Lett.* **125** 254502

[8] Sagaut P and Cambon C 2018 *Homogeneous Turbulence Dynamics* (Cham: Springer)

[9] Finlay C C 2008 Waves in the presence of magnetic fields, rotation and convection *Les Houches, Dynamos* ed P Cardin and L F Cugliandolo (Amsterdam: Elsevier) vol 88 pp 403–50

[10] Cattaneo F and Hughes D W 2006 Dynamo action in a rotating convective layer *J. Fluid Mech.* **553** 401–18

[11] Hughes D W and Cattaneo F 2008 The alpha-effect in rotating convection: size matters *J. Fluid Mech.* **594** 445–61

[12] Moffatt H K and Proctor M R E 1982 The role of the helicity spectrum function in turbulent dynamo theory *Geophys. Astrophys. Fluid Dyn.* **21** 265–83

[13] Brandenburg A and Subramanian K 2005 Astrophysical magnetic fields and nonlinear dynamo theory *Phys. Rep.* **417** 1–209

[14] Moffatt H K 2008 Magnetostrophic turbulence and the geodynamo *IUTAM Symposium on Computational Physics and New Perspectives in Turbulence, IUTAM Bookseries* ed Y Kaneda (Dordrecht: Springer) vol 4 pp 339–46

[15] Lesur G and Ogilvie G I 2008 Localized magnetorotational instability and its role in the accretion disc dynamo *Mon. Not. R. Astron. Soc.* **391** 1437–50

[16] Lesur G and Ogilvie G I 2008 On self-sustained dynamo cycles in accretion discs *Astron. Astrophys.* **488** 451–61

[17] Mizerski K A and Moffatt H K 2018 Dynamo generation of a magnetic field by decaying Lehnert waves in a highly conducting plasma *Geophys. Astrophys. Fluid Dyn.* **112** 165–74

[18] Moffatt H K 1970 Dynamo action associated with random inertial waves in a rotating conducting fluid *J. Fluid Mech.* **44** 705–19

[19] Ng C S and Bhattacharjee A 1996 Interaction of shear-Alfven wave packets: implication for weak magnetohydrodynamic turbulence in astrophysical plasmas *Astrophys. J.* **465** 845–54

[20] Aldridge K D and Lumb L I 1987 Inertial waves identified in the Earth's fluid outer core *Nature* **325** 421–3

[21] Braginsky S I 1967 Magnetic waves in the core of the Earth *Geomagn. Aeron.* **7** 851–9

[22] Braginsky S I 1980 Magnetic waves in the core of the Earth. II. *Geophys. Astrophys. Fluid Dyn.* **14** 189–208

[23] Hori K, Jones C A and Teed R J 2015 Slow magnetic Rossby waves in the Earth's core *Geophys. Res. Lett.* **42** 6622–9

[24] Yakhot V and Orszag S A 1986 Renormalization group analysis of turbulence. I. Basic theory *J. Sci. Comp* **1** 3–51

[25] Mizerski K A 2021 Renormalization group analysis of the magnetohydrodynamic turbulence and dynamo *J. Fluid Mech.* **926** A13

[26] Chandrasekhar S 1956 On force-free magnetic fields *Proc. Natl Acad. Sci. USA* **42** 1–5

[27] Chandrasekhar S and Kendall P C 1957 On force-free magnetic fields *Astrophys. J.* **126** 457–60

[28] Rädler K-H and Brandenburg A 2003 Contributions to the theory of a two-scale homogeneous dynamo experiment *Phys. Rev.* E **67** 026401

[29] Tassi E, Pegoraro F and Cicogna G 2008 Solutions and symmetries of force-free magnetic fields *Phys. Plasmas* **15** 092113

[30] Tobias S M 2021 The turbulent dynamo *J. Fluid Mech.* **912** P1

[31] Driscoll P and Olson P 2009 Polarity reversals in geodynamo models with core evolution *Earth Planet. Sci. Lett.* **282** 24–33

[32] Driscoll P and Olson P 2009 Effects of buoyancy and rotation on the polarity reversal frequency of gravitationally driven numerical dynamos *Geophys. J. Int.* **178** 1337–50

[33] Olson P 2013 The geodynamo's unique longevity *Phys. Today* **66** 30–5

[34] Mizerski K A 2021 Renormalization group analysis of the turbulent hydromagnetic dynamo: effect of anisotropy *Appl. Math. Comp.* **405** 126252

[35] Braginsky S I 1987 Waves in a stably stratified layer on the surface of the terrestrial core *Geomagn. Aeron.* **27** 410–4

[36] Braginsky S I 1999 Dynamics of the stably stratified ocean at the top of the core *Phys. Earth Planet. Int.* **111** 21–34

[37] Maxwell J C 1867 On the dynamical theory of gases *Phil. Trans. R. Soc. Lond.* **157** 49–88

[38] Krause F and Rädler K-H 1980 *Mean-Field Magnetohydrodynamics and Dynamo Theory* (Oxford: Pergamon)

[39] Rogachevskii I and Kleeorin N 2003 Electromotive force and large-scale magnetic dynamo in a turbulent flow with a mean shear *Phys. Rev.* E **68** 036301

[40] Rogachevskii I and Kleeorin N 2004 Nonlinear theory of a 'shear-current' effect and mean-field magnetic dynamos *Phys. Rev.* E **70** 046310

[41] Rogachevskii I, Kleeorin N and Liverts E 2006 Nonlinear shear-current dynamo and magnetic helicity transport in sheared turbulence *Geophys. Astrophys. Fluid Dyn.* **100** 537–57

[42] Yokoi N 2013 Cross helicity and related dynamo *Geophys. Astrophys. Fluid Dyn.* **107** 114–84

[43] Beck R, Brandenburg A, Moss D, Shukurov A and Sokoloff D 1996 Galactic magnetism: recent developments and perspectives *Annu. Rev. Astron. Astrophys.* **34** 155–206

[44] Schekochihin A A, Cowley S C, Taylor S F, Maron J L and McWilliams J C 2004 Simulations of the small-scale turbulent dynamo *Astrophys. J.* **612** 276–307

[45] Modjaz M, Moran J M, Kondratko P T and Greenhill L J 2005 Probing the magnetic field at subparsec radii in the accretion disk of NGC 4258 *Astrophys. J.* **626** 104–19

[46] Korpi M J, Brandenburg A, Shukurov A, Tuominen I and Nordlund Å 1999 A supernova-regulated interstellar medium: simulations of the turbulent multiphase medium *Astrophys. J.* **514** L99–102

IOP Publishing

Non-equilibrium Hydromagnetic Dynamos

Krzysztof A Mizerski

Chapter 4

The effect of nonlinear dynamics—weakly nonlinear renormalization group analysis

The investigation of turbulent magnetized flows is extremely difficult and requires sophisticated mathematical tools; see [1, 2] for a review of different approaches and results. The limit of weak turbulence considered in the previous chapter, was obtained by imposing a linear evolution of the turbulent fluctuations. In other words, the amplitude of turbulent fluctuations was assumed small enough to neglect the terms nonlinear in the fluctuations in the equations for their evolution. In such a way, the problem for evolution of fluctuations could be solved relatively easily and the obtained formulae for the fluctuating velocity and magnetic field used to calculate their nonlinear interactions (correlations) thus the electromotive force (EMF). As argued in [1], in some cases the regime of weak turbulence can be sustained for long times, nevertheless, it is much more common for natural systems to develop into the strong turbulence regime, where the evolution of turbulent fluctuations becomes nonlinear. Of course in such a case the evolution equations for fluctuations become very difficult to solve. We resolve this difficulty here, by applying the iterative renormalization technique, a statistical closure approximation which is based on systematic, subsequent (iterative) elimination of thin wave-number bands from the Fourier spectrum of rapidly evolving variables (cf e.g. [3]). The first to apply this technique in fluid mechanics were Wyld (1961) [4] and [5], to study the statistical properties of stationary, homogeneous and isotropic turbulence. However, the first to apply the renormalization group method in magnetohydrodynamics (MHD) was Moffatt (1981, 1983) [6, 7], who calculated the kinematic α coefficient, in a simplified case of a given stationary, homogeneous and isotropic turbulence flow field, unaffected by the Lorentz force. Later Yakhot and Orszag (1986) [8] published a very comprehensive work on renormalization of hydrodynamic equations in the absence of the magnetic field. A number of works on renormalization in non-magnetic stirred stationary turbulence followed, such as e.g. [9–14] with seminal reviews by Smith and Woodruff (1998) [15] and McComb (2014) [16]. It should also be noted, that Kleeorin and Rogachevskii [17] and Mizerski [18] have applied this method to MHD stationary, homogeneous and isotropic turbulence calculating the renormalized Lorentz force, α-effect and the turbulent magnetic diffusivity and viscosity.

In this chapter the iterative renormalization group method is applied to the full system of MHD equations, including the effect of back reaction of the Lorentz force on the flow which generates the magnetic field via dynamo action. The large-scale EMF is extracted from the dynamic nonlinear problem of non-stationary turbulence which is no longer assumed 'weak', i.e. the nonlinear terms in the evolution equations for turbulent fluctuations are now included. Thus we go beyond the weak turbulence regime and quantitatively express the effect of these nonlinearities on the dynamics of the large-scale flow. The turbulence is assumed to be driven by a small-scale, statistically random, homogeneous, isotropic and non-stationary forcing. The non-stationarity is proved to be important for the dynamo effect and its effect on the renormalized effective Lorentz force at large scales is also calculated. Finally, the mean-field equations are solved for a force-free mode.

However, the complications resulting from non-stationarity of the stirring force make full renormalization of the dynamical equations unattainable. Recent investigations of [19] involved application of the renormalization group method to study the effect of non-stationarity on the MHD turbulence in what could be called an 'intermediate regime of turbulence' or 'weakly nonlinear turbulence', in contrast to simple, linear, weak turbulence regime. Due to high complexity of the mathematical approach in the case of non-stationary turbulence the effect of nonlinear evolution of the fluctuations was included only at leading order in calculation of the effective turbulent EMF and the Lorentz force at each step of the iterative renormalization procedure; this corresponds to expansion of the full, renormalized transport coefficients up to first order in the Reynolds and magnetic Reynolds numbers. We adopt here the 'weakly nonlinear renormalization' approach, i.e. contrary to the full renormalization approach, at each step of the iterative procedure of a step-by-step elimination of thin wave number bands from the Fourier spectrum, only the terms at the leading order in the small Reynolds numbers are retained; in such a way the Taylor series in the Reynolds numbers cannot be contracted in the end, as is achieved in the full renormalization approaches, but instead we obtain weakly nonlinear expressions for the effective turbulent coefficients appearing in the final 'renormalized' equations for the large-scale fields, i.e. with the precision up to the second order terms in Re and Rm. These coefficients, i.e. the α coefficient and the turbulent coefficients describing the effective Lorentz force at large scales, contain the net nonlinear effect of weakly nonlinearly evolving short-wavelength fluctuations.

4.1 Dynamical equations

We now set $\boldsymbol{\Omega} = 0$, as we intend to study the isotropic case, and the helicity is injected by the small-scale statistically isotropic forcing defined by (3.132). The mean field equations are now

$$\frac{\partial\langle\mathbf{U}\rangle}{\partial t} + (\langle\mathbf{U}\rangle\cdot\nabla)\langle\mathbf{U}\rangle = -\nabla\langle\Pi\rangle + (\langle\mathbf{B}\rangle\cdot\nabla)\langle\mathbf{B}\rangle + \nu\nabla^2\langle\mathbf{U}\rangle - \nabla\cdot(\langle\mathbf{u}'\mathbf{u}'\rangle - \langle\mathbf{b}'\mathbf{b}'\rangle), \tag{4.1a}$$

$$\frac{\partial\langle\mathbf{B}\rangle}{\partial t} = \nabla\times(\langle\mathbf{U}\rangle\times\langle\mathbf{B}\rangle) + \nabla\times\langle\mathbf{u}'\times\mathbf{b}'\rangle + \eta\nabla^2\langle\mathbf{B}\rangle, \tag{4.1b}$$

$$\nabla\cdot\langle\mathbf{B}\rangle = 0, \quad \nabla\cdot\langle\mathbf{U}\rangle = 0. \tag{4.1c}$$

Next, since the calculation of the turbulent diffusivities will not be pursued we neglect the small gradients of the large-scale fields $\nabla\langle\mathbf{U}\rangle$, $\nabla\langle\mathbf{B}\rangle$, and the evolution of turbulent pulsations is governed by

$$\frac{\partial\mathbf{u}'}{\partial t} - \nu\nabla^2\mathbf{u}' + (\langle\mathbf{U}\rangle\cdot\nabla)\mathbf{u}' - (\langle\mathbf{B}\rangle\cdot\nabla)\mathbf{b}' + \nabla\Pi' = \mathbf{f} - \nabla\cdot(\mathbf{u}'\mathbf{u}' - \mathbf{b}'\mathbf{b}' - \langle\mathbf{u}'\mathbf{u}'\rangle + \langle\mathbf{b}'\mathbf{b}'\rangle), \tag{4.2a}$$

$$\frac{\partial\mathbf{b}'}{\partial t} - \eta\nabla^2\mathbf{b}' + (\langle\mathbf{U}\rangle\cdot\nabla)\mathbf{b}' - (\langle\mathbf{B}\rangle\cdot\nabla)\mathbf{u}' = \nabla\times(\mathbf{u}'\times\mathbf{b}' - \langle\mathbf{u}'\times\mathbf{b}'\rangle), \tag{4.2b}$$

$$\nabla\cdot\mathbf{b}' = 0, \quad \nabla\cdot\mathbf{u}' = 0. \tag{4.2c}$$

The aim of this analysis is to study the effect of the statistically homogeneous and isotropic but non-stationary and non-reflexionally symmetric (helical) forcing **f** in the Navier–Stokes equation on the large-scale dynamo process. The renormalization group technique is an iterative procedure based on successive elimination of thin wave-number bands from the Fourier spectrum of fluctuating fields. In this way the effect of thin bands of modes with shortest wavelengths on the remaining modes is calculated at each step of the procedure. The final aim of this approach is to obtain recursion equations for the α coefficient (mean EMF) and analogous coefficients for the turbulent Lorentz force (see equations (4.91a) and b)) as functions of the wave number. These recursion equations (provided in (4.82a–e)) are then solved for $k \to K_e$ in order to obtain the final forms of the large-scale EMF and the Lorentz force which appear in the renormalized mean-field equations and include the effect of nonlinear evolution of turbulent fluctuations.

4.2 The α-effect in weak turbulence with weak seed field

For the sake of a clear later reference we start by providing the formulae for the α coefficient when the turbulence is still weak and the amplitude of the mean magnetic field is small

$$\frac{\langle B\rangle^2}{\nu\eta k^2} \ll 1, \tag{4.3}$$

which corresponds to a problem of initial stage of the dynamo growth from a weak seed field. We can utilize the formulae (3.154a and b), (3.157a and b) and

$$
\begin{aligned}
&\frac{1}{|\gamma_u(\mathbf{q})|^2\,(\omega^2+\eta^2k^4)}\\
&=\frac{1}{\omega^4+\omega^2[k^4(\nu^2+\eta^2)-2(\mathbf{k}\cdot\langle\mathbf{B}\rangle)^2]+(\nu\eta k^4+(\mathbf{k}\cdot\langle\mathbf{B}\rangle)^2)^2}\\
&\approx\frac{1}{(\omega^2+\nu^2k^4)(\omega^2+\eta^2k^4)}+2(\mathbf{k}\cdot\langle\mathbf{B}\rangle)^2\frac{\omega^2-\nu\eta k^4}{(\omega^2+\nu^2k^4)^2(\omega^2+\eta^2k^4)^2},
\end{aligned}
\tag{4.4}
$$

which in the limit $t \ll 1/\tilde{\omega}$ and $\tilde{\omega} \ll \eta/\ell_e^2$ yield

$$
\bar{\alpha} = \bar{\alpha}_{\mathrm{st}} - \Gamma\bar{\alpha}_{\Delta\omega},
\tag{4.5}
$$

$$
\begin{aligned}
\bar{\alpha}_{\mathrm{st}} &= \frac{8\pi}{3}\eta D_1\langle B\rangle_i\int_{K_e}^{K}k\mathrm{d}k\int_{-\infty}^{\infty}\mathrm{d}\omega\frac{1}{(\omega^2+\nu^2k^4)(\omega^2+\eta^2k^4)}\\
&+\frac{16\pi}{5}\eta D_1\langle B\rangle^2\int_{K_e}^{K}k^3\mathrm{d}k\int_{-\infty}^{\infty}\mathrm{d}\omega\frac{\omega^2-\nu\eta k^4}{(\omega^2+\nu^2k^4)^2(\omega^2+\eta^2k^4)^2},
\end{aligned}
\tag{4.6}
$$

$$
\begin{aligned}
\bar{\alpha}_{\Delta\omega} = \frac{8\pi}{3}D_1\tilde{\omega}\int_{K_e}^{K}\frac{\mathrm{d}k}{k}\int_{-\infty}^{\infty}\mathrm{d}\omega&\left[\frac{\eta^2k^4-\omega^2}{(\omega^2+\nu^2k^4)(\omega^2+\eta^2k^4)^2}\right.\\
&\left.+\frac{\nu\eta k^4-\omega^2}{(\omega^2+\nu^2k^4)^2(\omega^2+\eta^2k^4)}\right]\\
-\frac{8\pi}{5}D_1\tilde{\omega}\langle B\rangle^2\int_{K_e}^{K}k\mathrm{d}k\int_{-\infty}^{\infty}\mathrm{d}\omega&\left[\frac{4(\omega^2-\nu\eta k^4)^2}{(\omega^2+\nu^2k^4)^3(\omega^2+\eta^2k^4)^2}\right.\\
&\left.+\frac{2\omega^4-\eta(5\nu+7\eta)k^4\omega^2+\eta^3k^8(3\nu-\eta)}{(\omega^2+\nu^2k^4)^2(\omega^2+\eta^2k^4)^3}\right],
\end{aligned}
\tag{4.7}
$$

where K denotes a standard cut-off of the k-spectrum (either the viscous or the resistive one, whichever is smaller) and, in principle, can be infinite. Performing the ω integration one obtains

$$
\begin{aligned}
\bar{\alpha}_A &= \frac{8\pi^2}{3}\frac{D_1}{\nu(\nu+\eta)}\int_{K_e}^{K}\frac{\mathrm{d}k}{k^5}-\frac{4\pi^2\Gamma}{3}\frac{D_1\tilde{\omega}(3\nu+\eta)}{\nu^2(\nu+\eta)^2}\int_{K_e}^{K}\frac{\mathrm{d}k}{k^7}\\
&\overset{K_e\ll K}{\approx}\frac{2\pi^2}{3}\frac{D_1}{\nu(\nu+\eta)}\frac{1}{K_e^4}-\frac{2\pi^2\Gamma}{9}\frac{D_1\tilde{\omega}(3\nu+\eta)}{\nu^2(\nu+\eta)^2}\frac{1}{K_e^6},
\end{aligned}
\tag{4.8}
$$

$$
\begin{aligned}
\bar{\alpha}_S &= \frac{8\pi^2}{5}\frac{D_1}{\nu^2\eta(\nu+\eta)}\int_{K_e}^{K}\frac{\mathrm{d}k}{k^7}-\frac{4\pi^2\Gamma}{5}\frac{D_1\tilde{\omega}(\nu^2+5\nu\eta+2\eta^2)}{\nu^3\eta^2(\nu+\eta)^2}\int_{K_e}^{K}\frac{\mathrm{d}k}{k^9}\\
&\overset{K_e\ll K}{\approx}\frac{4\pi^2}{15}\frac{D_1}{\nu^2\eta(\nu+\eta)}\frac{1}{K_e^6}-\frac{\pi^2\Gamma}{10}\frac{D_1\tilde{\omega}(\nu^2+5\nu\eta+2\eta^2)}{\nu^3\eta^2(\nu+\eta)^2}\frac{1}{K_e^8},
\end{aligned}
\tag{4.9}
$$

where the coefficients $\bar{\alpha}_A$ and $\bar{\alpha}_S$ are defined in the following way

$$
\bar{\alpha} = \bar{\alpha}_A - \bar{\alpha}_S\langle B\rangle^2.
\tag{4.10}
$$

The $\bar{\alpha}_S$ coefficient includes the effect of the Lorentz force and therefore it is mainly responsible for saturation of the large-scale field, whereas the coefficient $\bar{\alpha}_A$ leads to amplification (hence the subscripts).

It is worth mentioning that in principle the non-equilibrium contributions to the α-effect survive even when $\eta = 0$, although, as explained above and also later in section 4.3.2, the exact form obtained in (4.8 and 4.9) corresponds only to the case $\eta \neq 0$ (cf 4.89), since in the asymptotic limit considered in this chapter the diffusivities are needed for regularization of the Fourier ω-integrals. However, the non-equilibrium α-effect does not crucially depend on the magnetic diffusivity and at

$$\eta = 0, \tag{4.11}$$

(with $\Gamma = 1$) under some additional regularizing assumptions such as (cf equation (3.5) in [20])

$$D_0 = D_0(\omega) \sim \omega^2, \quad D_1 = D_1(\omega) \sim \omega^2 \text{ when } \omega \to 0, \tag{4.12}$$

the total α-effect consists solely of the non-equilibrium contributions

$$\begin{aligned}\bar{\alpha} = \bar{\alpha}_{\Delta\omega} = &- \frac{8\pi}{3}\bar{\varpi}\int_{K_e}^{K}\frac{\mathrm{d}k}{k}\int_{-\infty}^{\infty}\mathrm{d}\omega\frac{D_1(\omega)(2\omega^2 + \nu^2k^4)}{(\omega^2 + \nu^2k^4)^2\omega^2} \\ &- \frac{16\pi}{5}\bar{\varpi}\langle B\rangle^2\int_{K_e}^{K}k\mathrm{d}k\int_{-\infty}^{\infty}\mathrm{d}\omega\frac{D_1(\omega)(3\omega^2 + \nu^2k^4)}{(\omega^2 + \nu^2k^4)^3\omega^2},\end{aligned} \tag{4.13}$$

and $\bar{\alpha}_{\mathrm{st}} = 0$.

4.2.1 Energy time evolution in weak turbulence

For the problem of statistically non-stationary turbulence it is also instructive to provide explicit time evolution laws for the turbulent kinetic and magnetic energies, $\langle e_k\rangle = \langle \mathbf{u}^{'2}\rangle$, $\langle e_m\rangle = \langle \mathbf{b}^{'2}\rangle$, which by the use of (3.152) take the following form

$$\begin{aligned}\frac{\partial\langle e_m\rangle}{\partial t} = &-\langle B\rangle_m\langle B\rangle_n\int \mathrm{d}^4q\int \mathrm{d}^4q'\frac{k_m k_n' \mathrm{e}^{\mathrm{i}[(\mathbf{k}+\mathbf{k}')\cdot\mathbf{x}-(\omega+\omega')t]}}{\gamma_u(\mathbf{q}')\gamma_u(\mathbf{q})\gamma_\eta(\mathbf{q}')}\left\langle \hat{f}_i(\mathbf{q})\hat{f}_i(\mathbf{q}')\right\rangle \\ &+\eta\langle B\rangle_m\langle B\rangle_n\int \mathrm{d}^4q\int \mathrm{d}^4q'\frac{k_m k_n' k^{'2} \mathrm{e}^{\mathrm{i}[(\mathbf{k}+\mathbf{k}')\cdot\mathbf{x}-(\omega+\omega')t]}}{\gamma_u(\mathbf{q}')\gamma_u(\mathbf{q})\gamma_\eta(\mathbf{q}')\gamma_\eta(\mathbf{q})}\left\langle \hat{f}_i(\mathbf{q})\hat{f}_i(\mathbf{q}')\right\rangle,\end{aligned} \tag{4.14}$$

$$\begin{aligned}\frac{\partial\langle e_k\rangle}{\partial t} = &\int \mathrm{d}^4q\int \mathrm{d}^4q'\frac{1}{\gamma_u(\mathbf{q})}\left\langle \hat{f}_i(\mathbf{q})\hat{f}_i(\mathbf{q}')\right\rangle \mathrm{e}^{\mathrm{i}[(\mathbf{k}+\mathbf{k}')\cdot\mathbf{x}-(\omega+\omega')t]} \\ &-\langle B\rangle_m\langle B\rangle_n\int \mathrm{d}^4q\int \mathrm{d}^4q'\frac{k_m' k_n' \mathrm{e}^{\mathrm{i}[(\mathbf{k}+\mathbf{k}')\cdot\mathbf{x}-(\omega+\omega')t]}}{\gamma_u(\mathbf{q}')\gamma_u(\mathbf{q})\gamma_\eta(\mathbf{q}')}\left\langle \hat{f}_i(\mathbf{q})\hat{f}_i(\mathbf{q}')\right\rangle \\ &-\nu\int \mathrm{d}^4q\int \mathrm{d}^4q'\frac{k^{'2}}{\gamma_u(\mathbf{q}')\gamma_u(\mathbf{q})}\left\langle \hat{f}_i(\mathbf{q})\hat{f}_i(\mathbf{q}')\right\rangle \mathrm{e}^{\mathrm{i}[(\mathbf{k}+\mathbf{k}')\cdot\mathbf{x}-(\omega+\omega')t]}.\end{aligned} \tag{4.15}$$

On introducing the explicit expressions for the γ-coefficients $\gamma_\nu = -\mathrm{i}\omega + \nu k^2$, $\gamma_\eta = -\mathrm{i}\omega + \eta k^2$, $\gamma_u = \gamma_\nu + (\mathbf{k}\cdot\langle\mathbf{B}\rangle)^2/\gamma_\eta$ and the force correlations (3.132) in the limit $t \ll 1/\tilde{\omega}$ and $\tilde{\omega} \ll \eta/\ell_e^2$ we can calculate the integrals in (4.14) and (4.15), which yields

$$\begin{aligned}\frac{\partial\langle e_m\rangle}{\partial t} &= \frac{4\pi^2}{3}\frac{D_0\langle B\rangle^2}{\nu(\nu+\eta)K_e^2}\left[1 - \frac{3\nu+\eta}{\nu(\nu+\eta)}\frac{\Gamma\tilde{\omega}}{4K_e^2}\right] \\ &- \frac{4\pi^2}{3}\frac{D_0\langle B\rangle^2}{\nu(\nu+\eta)K_e^2}\left[1 - \frac{\nu^2+4\nu\eta+\eta^2}{\nu\eta(\nu+\eta)}\frac{\Gamma\tilde{\omega}}{4K_e^2}\right] \\ &= \frac{\pi^2\Gamma\tilde{\omega}D_0\langle B\rangle^2}{3\nu\eta(\nu+\eta)K_e^4}.\end{aligned} \tag{4.16}$$

The kinetic energy evolution law can also be obtained in an explicit form

$$\begin{aligned}\frac{\partial\langle e_k\rangle}{\partial t} &= 8\pi^2 D_0 \ln\frac{K_\nu}{K_e} - \frac{4\pi^2}{3}\frac{D_0\langle B\rangle^2}{\nu(\nu+\eta)K_e^2}\left[1 - \frac{3\nu+\eta}{\nu(\nu+\eta)}\frac{\Gamma\tilde{\omega}}{4K_e^2}\right] \\ &- 8\pi^2 D_0\left(\ln\frac{K_\nu}{K_e} - \frac{\Gamma\tilde{\omega}}{4\nu K_e^2}\right) + \frac{4\pi^2}{3}\frac{D_0\langle B\rangle^2}{\nu(\nu+\eta)K_e^2}\left[1 - \frac{2\nu+\eta}{\nu(\nu+\eta)}\frac{\Gamma\tilde{\omega}}{2K_e^2}\right] \\ &= \frac{2\pi^2 D_0\Gamma\tilde{\omega}}{\nu K_e^2} - \frac{\pi^2\Gamma\tilde{\omega}D_0\langle B\rangle^2}{3\nu^2(\nu+\eta)K_e^4}.\end{aligned} \tag{4.17}$$

The net loss of the kinetic energy is described by the term

$$-\frac{\pi^2\Gamma\tilde{\omega}D_0\langle B\rangle^2}{3\nu^2(\nu+\eta)K_e^4}, \tag{4.18}$$

and results from viscous dissipation and the energy transfer $\langle e_k\rangle \rightarrow \langle e_m\rangle$ via the Lorentz force. The net gain of the magnetic energy

$$\frac{\pi^2\Gamma\tilde{\omega}D_0\langle B\rangle^2}{3\nu\eta(\nu+\eta)K_e^4}, \tag{4.19}$$

results from the Lorentz force transfer from $\langle e_k\rangle$ reduced by the effect of resistive dissipation. Symbolically we can write

$$\frac{\partial\langle e_m\rangle}{\partial t} = \mathbb{L}_{K\to M} - \mathbb{D}_\eta, \qquad \frac{\partial\langle e_k\rangle}{\partial t} = \mathbb{F} - \mathbb{L}_{K\to M} - \mathbb{D}_{\nu 0} + \mathbb{D}_{\nu L}, \tag{4.20}$$

where

$$\mathbb{L}_{K\to M} = \frac{4\pi^2}{3}\frac{D_0\langle B\rangle^2}{\nu(\nu+\eta)K_e^2}\left[1 - \frac{3\nu+\eta}{\nu(\nu+\eta)}\frac{\Gamma\tilde{\omega}}{4K_e^2}\right], \tag{4.21}$$

denotes the energy transfer $\langle e_k \rangle \to \langle e_m \rangle$ via the Lorentz force,

$$\mathbb{D}_\eta = \frac{4\pi^2}{3}\frac{D_0\langle B\rangle^2}{\nu(\nu+\eta)K_e^2}\left[1 - \frac{\nu^2+4\nu\eta+\eta^2}{\nu\eta(\nu+\eta)}\frac{\Gamma\tilde{\omega}}{4K_e^2}\right], \tag{4.22}$$

denotes the magnitude of the Ohmic energy losses,

$$\mathbb{D}_{\nu 0} = 8\pi^2 D_0\left(\ln\frac{K_\nu}{K_e} - \frac{\Gamma\tilde{\omega}}{4\nu K_e^2}\right), \quad \mathbb{D}_{\nu L} = \frac{4\pi^2}{3}\frac{D_0\langle B\rangle^2}{\nu(\nu+\eta)K_e^2}\left[1 - \frac{2\nu+\eta}{\nu(\nu+\eta)}\frac{\Gamma\tilde{\omega}}{2K_e^2}\right], \tag{4.23}$$

describe the viscous energy losses which are reduced by action of the Lorentz force; the energy losses uninfluenced by the Lorentz force are denoted by $\mathbb{D}_{\nu 0}$ and the diminishing effect of the Lorentz force is denoted by $\mathbb{D}_{\nu L}$ (in stationary case $\mathbb{D}_{\nu L} = \mathbb{D}_\eta$). Finally, the source (forcing) term is given by

$$\mathbb{F} = 8\pi^2 D_0 \ln\frac{K_\nu}{K_e}. \tag{4.24}$$

Naturally, all the terms include the effect of non-stationarity, which can act either to quench or boost them, depending on the sign of $\Gamma\tilde{\omega}$, that is on whether the mean energy grows or decays. We also emphasize that within the considered limits (4.3) and $t \ll 1/\tilde{\omega}$ and $\tilde{\omega} \ll \eta/\ell_e^2$, ν/ℓ_e^2 the corrections from the Lorentz force and from non-stationarity are small and $\langle e_m \rangle \ll \langle e_k \rangle$. The total energy satisfies

$$\frac{\partial}{\partial t}(\langle e_k \rangle + \langle e_m \rangle) = (\mathbb{F} - \mathbb{D}_{\nu 0}) + (\mathbb{D}_{\nu L} - \mathbb{D}_\eta), \tag{4.25}$$

where of course the entire right hand side is proportional to $\Gamma\tilde{\omega}$. Note, that at $Pm = \nu/\eta = 1$ even in the non-stationary case we get $\mathbb{D}_{\nu L} = \mathbb{D}_\eta$ (in the stationary case this is true for all values of Pm).

4.3 Renormalization procedure of the MHD equations

Let us introduce non-dimensional variables

$$\mathbf{f} \mapsto \mathcal{F}\mathbf{f}, \ \mathbf{U} \mapsto \mathcal{U}\mathbf{U}, \ \mathbf{b}' \mapsto \mathcal{B}\mathbf{b}', \ \langle\mathbf{B}\rangle \mapsto B_0\langle\mathbf{B}\rangle, \quad \Pi \mapsto \ell\mathcal{F}\Pi, \tag{4.26a}$$

$$\mathbf{x} \mapsto \ell\mathbf{x}, \ t \mapsto \frac{\mathcal{U}}{\mathcal{F}}t \tag{4.26b}$$

$$D_0 \mapsto \mathcal{F}\mathcal{U}D_0, \ D_1 \mapsto \frac{\mathcal{F}\mathcal{U}}{\ell}D_1, \tag{4.26c}$$

where ℓ is some length scale of variation of fluctuations,

$$\frac{2\pi}{K_e} \leqslant \ell \leqslant \frac{2\pi}{K}, \qquad K = \min(K_\nu, K_\eta), \tag{4.27}$$

and the scales $\mathcal{U}$ and $\mathcal{B}$ are defined by the norms of the initial fluctuation fields, i.e. $\mathcal{U} = \| \mathbf{u}(t = 0)\|_{L^2}$ and $\mathcal{B} = \| \mathbf{b}(t = 0)\|_{L^2}$; in an analogous way we define $B_0 = \| \langle \mathbf{B} \rangle(t = 0)\|_{L^2}$ and the characteristic length-scale of variation of the mean fields will be denoted by

$$L \gg \ell_e = \frac{2\pi}{K_e}. \tag{4.28}$$

The scale of the stirring force $\mathcal{F}$ can be chosen later if needed, and here we introduce the following non-dimensional parameters

$$Ro = \frac{\mathcal{U}^2}{\ell \mathcal{F}}, \quad E_v = \frac{\nu \mathcal{U}}{\ell^2 \mathcal{F}}, \quad E_\eta = \frac{\eta \mathcal{U}}{\ell^2 \mathcal{F}}, \quad \tilde{\beta} = \frac{B_0}{\mathcal{B}}, \tag{4.29}$$

and

$$H = \frac{\mathcal{B}^2}{\mathcal{U}^2}, \quad \varepsilon = \frac{\ell}{L}, \tag{4.30}$$

where the names Ro, E_ν, E_η correspond to standard Rossby and Ekman numbers, if one conceptually associates the helical part of the forcing with the Coriolis force; however, we want to avoid the effect of background rotation here, as it is strongly anisotropic, bringing unnecessary complications to the calculation, for which the major aim is the demonstration of the nonlinear non-equilibrium effect. Defining

$$\mathbf{X} = \varepsilon \mathbf{x}, \qquad \bar{\nabla} = \nabla_{\mathbf{X}}, \tag{4.31}$$

we can rewrite the dynamical equations in non-dimensional form

$$\begin{aligned} \frac{\partial \langle \mathbf{U} \rangle}{\partial t} + \varepsilon Ro (\langle \mathbf{U} \rangle \cdot \bar{\nabla}) \langle \mathbf{U} \rangle = &- \varepsilon \bar{\nabla} \langle \Pi \rangle + \varepsilon Ro \tilde{\beta}^2 H (\langle \mathbf{B} \rangle \cdot \bar{\nabla}) \langle \mathbf{B} \rangle + E_\nu \varepsilon^2 \bar{\nabla}^2 \langle \mathbf{U} \rangle \\ &- \varepsilon Ro (\bar{\nabla} \cdot \langle \mathbf{u}' \mathbf{u}' \rangle - H \bar{\nabla} \cdot \langle \mathbf{b}' \mathbf{b}' \rangle), \end{aligned} \tag{4.32a}$$

$$\frac{\partial \langle \mathbf{B} \rangle}{\partial t} = \varepsilon Ro \bar{\nabla} \times (\langle \mathbf{U} \rangle \times \langle \mathbf{B} \rangle) + \varepsilon Ro \tilde{\beta}^{-1} \bar{\nabla} \times \langle \mathbf{u}' \times \mathbf{b}' \rangle + \varepsilon^2 E_\eta \bar{\nabla}^2 \langle \mathbf{B} \rangle, \tag{4.32b}$$

$$\begin{aligned} &\frac{\partial \mathbf{u}'}{\partial t} - E_\nu \nabla^2 \mathbf{u}' + Ro (\langle \mathbf{U} \rangle \cdot \nabla) \mathbf{u}' - Ro \tilde{\beta} H (\langle \mathbf{B} \rangle \cdot \nabla) \mathbf{b}' + \nabla \Pi' \\ &\quad = \mathbf{f} - Ro \nabla \cdot (\mathbf{u}'\mathbf{u}' - H \mathbf{b}'\mathbf{b}' - \langle \mathbf{u}'\mathbf{u}' \rangle + H \langle \mathbf{b}'\mathbf{b}' \rangle), \end{aligned} \tag{4.32c}$$

$$\begin{aligned} &\frac{\partial \mathbf{b}'}{\partial t} - E_\eta \nabla^2 \mathbf{b}' + Ro (\langle \mathbf{U} \rangle \cdot \nabla) \mathbf{b}' \\ &\quad = Ro \tilde{\beta} (\langle \mathbf{B} \rangle \cdot \nabla) \mathbf{u}' + Ro \nabla \times (\mathbf{u}' \times \mathbf{b}' - \langle \mathbf{u}' \times \mathbf{b}' \rangle), \end{aligned} \tag{4.32d}$$

$$\nabla \cdot \langle \mathbf{B} \rangle = 0, \quad \nabla \cdot \langle \mathbf{U} \rangle = 0, \quad \nabla \cdot \mathbf{b}' = 0, \quad \nabla \cdot \mathbf{u}' = 0. \tag{4.32e}$$

Note, that $H \neq 0$ marks the presence of the Lorentz force, i.e. when $H = 0$ the problem becomes kinematic. We now introduce the following assumptions

$$Ro \ll 1, \quad 1 \ll \tilde{\beta} \ll Ro^{-1}, \tag{4.33a}$$

$$\frac{\kappa_m}{K_e} \sim \frac{\ell_e}{L} \ll Ro^2 \Rightarrow \varepsilon \ll Ro^2, \tag{4.33b}$$

and

$$H = \mathcal{O}(1), \tag{4.34}$$

where the assumption (4.33b) is required for consistency with neglection of the gradients of means in the equations for fluctuations. Using the short notation (3.83) the equations for fluctuations in the Fourier space take the form

$$\begin{aligned} &\left[-\mathrm{i}(\omega - Ro\mathbf{k}\cdot\langle\mathbf{U}\rangle) + E_\nu k^2\right]\hat{u}_i(\mathbf{q}) - \mathrm{i}Ro\tilde{\beta}H(\mathbf{k}\cdot\langle\mathbf{B}\rangle)\hat{b}_i(\mathbf{q}) + \mathrm{i}k_i\hat{\Pi} \\ &= \hat{f}_i(\mathbf{q}) - \mathrm{i}Rok_j\left[\mathbb{I}_{ij}^{(u)} - H\mathbb{I}_{ij}^{(b)}\right] + \mathrm{i}Rok_j\left[\left\langle\mathbb{I}_{ij}^{(u)}\right\rangle - H\left\langle\mathbb{I}_{ij}^{(b)}\right\rangle\right], \end{aligned} \tag{4.35a}$$

$$\begin{aligned} &\left[-\mathrm{i}(\omega - Ro\mathbf{k}\cdot\langle\mathbf{U}\rangle) + E_\eta k^2\right]\hat{b}_i(\mathbf{q}) \\ &= \mathrm{i}Ro\tilde{\beta}(\mathbf{k}\cdot\langle\mathbf{B}\rangle)\hat{u}_i(\mathbf{q}) + \mathrm{i}Rok_j\mathbb{I}_{ij}^{(ub)} - \mathrm{i}Rok_j\left\langle\mathbb{I}_{ij}^{(ub)}\right\rangle, \end{aligned} \tag{4.35b}$$

$$\mathbf{k}\cdot\hat{\mathbf{u}}(\mathbf{q}) = 0, \quad \mathbf{k}\cdot\hat{\mathbf{b}}(\mathbf{q}) = 0, \quad \mathbf{k}\cdot\hat{\mathbf{f}}(\mathbf{q}) = 0, \tag{4.35c}$$

where

$$\mathbb{I}_{ij}^{(u)} = \theta_{K_\nu}\int_{K_e}^{K_\nu} d^4q'\hat{u}_i(\mathbf{q}-\mathbf{q}')\hat{u}_j(\mathbf{q}'), \qquad \mathbb{I}_{ij}^{(b)} = \theta_{K_\eta}\int_{K_e}^{K_\eta} d^4q'\hat{b}_i(\mathbf{q}-\mathbf{q}')\hat{b}_j(\mathbf{q}'), \tag{4.36}$$

$$\mathbb{I}_{ij}^{(ub)} = \epsilon_{ijk}\epsilon_{kmn}\theta_K\int_{K_e}^{K} d^4q'\hat{u}_m(\mathbf{q}-\mathbf{q}')\hat{b}_n(\mathbf{q}'), \tag{4.37}$$

with

$$K = \min\left(K_\nu, K_\eta\right), \qquad \theta_{K_i} = \theta(K_i - k). \tag{4.38}$$

Factors θ were added for clarity in the following calculations. Since the mean velocity $\langle\mathbf{U}\rangle$ only creates a shift of the frequency $\omega \to \omega - \mathbf{k}\cdot\langle\mathbf{U}\rangle$, in the following we will simply absorb it into the frequency. This is allowed here, since we do not introduce into the problem the time-scale separation and the Fourier frequencies take all real values from $-\infty$ to ∞, whereas the norm of the wave vectors is bounded from above (cf also (3.6a–c) and the short discussion below). The convolution integrals possess the following symmetry properties

$$\mathbb{I}_{ij}^{(u)} = \mathbb{I}_{ji}^{(u)}, \qquad \mathbb{I}_{ij}^{(b)} = \mathbb{I}_{ji}^{(b)}, \qquad \mathbb{I}_{ij}^{(ub)} = -\mathbb{I}_{ji}^{(ub)}. \tag{4.39}$$

It should be noted at this stage, that the convolution integrals, which represent the nonlinear interactions between fluctuating turbulent fields are *not* neglected in the evolutional equations for fluctuations (4.35a and b), and their role is crucial. In fact it is the aim of the entire renormalization procedure to go beyond the weak turbulence regime and quantitatively express the effect of those nonlinearities on the dynamics of the mean fields. Therefore, contrary to the previous sections, where the mean dynamo effect was studied in the regime of weak turbulence (i.e. under neglection of $\mathbb{I}_{ij}^{(u)}$, $\mathbb{I}_{ij}^{(b)}$ and $\mathbb{I}_{ij}^{(ub)}$ in (4.35a and b)), we will now investigate the effect of nonlinear evolution of the turbulent fluctuations on the mean EMF.

In order to eliminate pressure we apply the projection operator

$$P_{ij}(\mathbf{k}) = \delta_{ij} - \frac{k_i k_j}{k^2}, \tag{4.40}$$

to both sides of the Navier–Stokes equation (4.35a) and after some simple algebra we get explicit expressions for $\hat{u}_i(\mathbf{q})$ and $\hat{b}_i(\mathbf{q})$

$$\begin{aligned} \hat{u}_i(\mathbf{q}) = \frac{1}{\gamma_u}\hat{f}_i(\mathbf{q}) - \frac{Ro}{\gamma_u}\bigg\{ & \frac{1}{2}\mathrm{i}P_{imn}(\mathbf{k})[\mathbb{I}_{mn}^{(u)} - H\mathbb{I}_{mn}^{(b)} - \langle \mathbb{I}_{mn}^{(u)}\rangle + H\langle \mathbb{I}_{mn}^{(b)}\rangle] \\ & + HRo\tilde{\beta}\frac{\mathbf{k}\cdot\langle\mathbf{B}\rangle}{\gamma_\eta}k_j\left[\mathbb{I}_{ij}^{(ub)} - \left\langle \mathbb{I}_{ij}^{(ub)}\right\rangle\right]\bigg\}, \end{aligned} \tag{4.41a}$$

$$\begin{aligned} \hat{b}_i(\mathbf{q}) = {} & \mathrm{i}Ro\tilde{\beta}\frac{\mathbf{k}\cdot\langle\mathbf{B}\rangle}{\gamma_u\gamma_\eta}\hat{f}_i(\mathbf{q}) \\ & + \mathrm{i}\frac{Ro}{\gamma_\eta}\left[1 - H\left(Ro\tilde{\beta}\right)^2\frac{(\mathbf{k}\cdot\langle\mathbf{B}\rangle)^2}{\gamma_u\gamma_\eta}\right]k_j\left[\mathbb{I}_{ij}^{(ub)} - \left\langle \mathbb{I}_{ij}^{(u^< b^<)}\right\rangle\right] \\ & + \frac{1}{2}Ro\left(Ro\tilde{\beta}\right)\frac{\mathbf{k}\cdot\langle\mathbf{B}\rangle}{\gamma_u\gamma_\eta}P_{imn}(\mathbf{k})\left[\mathbb{I}_{mn}^{(u)} - H\mathbb{I}_{mn}^{(b)} - \left\langle \mathbb{I}_{mn}^{(u)}\right\rangle + H\left\langle \mathbb{I}_{mn}^{(b)}\right\rangle\right], \end{aligned} \tag{4.41b}$$

where

$$P_{imn}(\mathbf{k}) = k_m P_{in}(\mathbf{k}) + k_n P_{im}(\mathbf{k}), \tag{4.42}$$

$$\gamma_u = -\mathrm{i}\omega + E_\nu k^2 + H\left(Ro\tilde{\beta}\right)^2\frac{(\mathbf{k}\cdot\langle\mathbf{B}\rangle)^2}{\gamma_\eta}, \quad \gamma_\eta = -\mathrm{i}\omega + E_\eta k^2, \tag{4.43}$$

and, as mentioned above, the constant term $-Ro\mathbf{k}\cdot\langle\mathbf{U}\rangle$ has been absorbed into the frequency ω. The following simple transformation allows us to return to original dimensional variables

$$E_\nu \to \nu, \ E_\eta \to \eta, \ H \to 1, \ Ro \to 1, \ \tilde{\beta} \to 1. \tag{4.44}$$

In the non-dimensional variables the isotropic and homogeneous, but *non-stationary* forcing is still defined by the same formula

$$\left\langle \hat{f}_i(\mathbf{k},\omega)\hat{f}_j(\mathbf{k}',\omega')\right\rangle = \left[\frac{D_0}{k^3}P_{ij}(\mathbf{k}) + \mathrm{i}\frac{D_1}{k^5}\epsilon_{ijk}k_k\right]\delta(\mathbf{k}+\mathbf{k}') \times \left[\delta(\omega+\omega') + \frac{\Gamma}{2\mathrm{i}}(\delta(\omega+\omega'+\tilde{\omega}) - \delta(\omega+\omega'-\tilde{\omega}))\right], \tag{4.45}$$

so that for $t' = t$

$$\left\langle f_i(\mathbf{x},t)f_i(\mathbf{x}',t)\right\rangle \sim 1 + \Gamma\sin(\tilde{\omega}t). \tag{4.46}$$

(of course when we return to dimensional variables $D_0 \to D_0$ and $D_1 \to D_1$).

We can now comment on the second assumption in (4.33a), which states, that the mean field is much stronger than the fluctuating one, but $Ro\tilde{\beta} \ll 1$. The former allows for proper formulation of the problem, since at strong mean field the fluctuational magnetic field (4.41b) is expressed at leading order by the stirring force, and the nonlinearities, which are of the order $\mathcal{O}(Ro)$, can be treated in perturbational sense. The iterational procedure of renormalization is then applicable. On the other hand, the assumption $Ro\tilde{\beta} \ll 1$ allows for expansion of the γ_u^{-1} factors, and therefore explicit calculation of Fourier integrals in the nonlinear $\mathbb{I}$-terms. In turn, the final recursion differential relations for the coefficients describing the renormalized EMF and Lorentz force can be solved analytically, thus the full leading order form of the large-scale EMF and the effective large-scale Lorentz force can be determined.

We now start the renormalization procedure of taking successive little bites off the Fourier spectrum from the short-wavelength side in order to obtain the final nonlinear effect of the fluctuations on the means. At the first step of the procedure we introduce a parameter k_1, which satisfies

$$\delta k = K_{\max} - k_1 \ll 1, \quad \text{where} \quad K_{\max} = \max\left(K_\nu, K_\eta\right), \tag{4.47}$$

and divide the Fourier spectrum into two parts

$$\hat{u}_i^>(\mathbf{k},\omega) = \theta(k - k_1)\hat{u}_i(\mathbf{k},\omega), \qquad (\text{for } k_1 < k < K_{\max}), \tag{4.48}$$

$$\hat{u}_i^<(\mathbf{k},\omega) = \theta(k_1 - k)\hat{u}_i(\mathbf{k},\omega), \qquad (\text{for } k < k_1), \tag{4.49}$$

and the same way for $\hat{\mathbf{b}}$ and $\hat{\mathbf{f}}$. The equations for the fields $\hat{u}_i^<(\mathbf{k},\omega)$ and $\hat{b}_i^<(\mathbf{k},\omega)$ are obtained by projecting (4.35a) on the direction perpendicular to $\mathbf{k}$ (with the use of (4.40)), and averaging both $\mathbf{P}(\mathbf{k})\cdot$ (4.35a) and (4.35b) over the first shell $(k_1 < k < K_{\max})$

$$\begin{aligned}(-\mathrm{i}\omega + E_\nu k^2)\hat{u}_i^<(\mathbf{q}) &= \hat{f}_i^<(\mathbf{q}) + \mathrm{i}Ro\tilde{\beta}H(\mathbf{k}\cdot\langle\mathbf{B}\rangle)\hat{b}_i^<(\mathbf{q}) \\ &- \frac{1}{2}\mathrm{i}RoP_{imn}(\mathbf{k})[\mathbb{I}_{mn}^{(u<)} - H\mathbb{I}_{mn}^{(b<)} - \langle\mathbb{I}_{mn}^{(u<)}\rangle + H\langle\mathbb{I}_{mn}^{(b<)}\rangle] \\ &- \frac{1}{2}\mathrm{i}RoP_{imn}(\mathbf{k})\Bigg\{\theta_{K_\nu}\int_{K_e}^{K_\nu} d^4q'\left[\left\langle\hat{u}_m^>(\mathbf{q}')\hat{u}_n^>(\mathbf{q}-\mathbf{q}')\right\rangle_c - \left\langle\hat{u}_m^>(\mathbf{q}')\hat{u}_n^>(\mathbf{q}-\mathbf{q}')\right\rangle\right] \\ &\quad - H\theta_{K_\eta}\int_{K_e}^{K_\eta} d^4q'\left[\left\langle\hat{b}_m^>(\mathbf{q}')\hat{b}_n^>(\mathbf{q}-\mathbf{q}')\right\rangle_c - \left\langle\hat{b}_m^>(\mathbf{q}')\hat{b}_n^>(\mathbf{q}-\mathbf{q}')\right\rangle\right]\Bigg\},\end{aligned} \tag{4.50a}$$

$$\left(-\mathrm{i}\omega + E_\eta k^2\right)\hat{b}_i^<(\mathbf{q}) = \mathrm{i}Ro\tilde{\beta}(\mathbf{k}\cdot\langle\mathbf{B}\rangle)\hat{u}_i^<(\mathbf{q}) + \mathrm{i}Rok_j\left[\mathbb{I}_{ij}^{(u^<b^<)} - \left\langle\mathbb{I}_{ij}^{(u^<b^<)}\right\rangle\right]$$
$$+\,\mathrm{i}Rok_j\epsilon_{ijk}\epsilon_{kmn}\theta_K\int_{K_e}^{K} d^4q'\left[\left\langle\hat{u}_m^>(\mathbf{q}')\hat{b}_n^>(\mathbf{q}-\mathbf{q}')\right\rangle_c - \left\langle\hat{u}_m^>(\mathbf{q}')\hat{b}_n^>(\mathbf{q}-\mathbf{q}')\right\rangle\right]. \tag{4.50b}$$

On the other hand, to get equations for $\hat{u}_i^>(\mathbf{k},\omega)$ and $\hat{b}_i^>(\mathbf{k},\omega)$ we utilize (4.41a and b) which yields

$$\hat{u}_i^>(\mathbf{q}) = \frac{1}{\gamma_u}\hat{f}_i^>(\mathbf{q}) - \frac{1}{2}\mathrm{i}Ro\frac{P_{imn}(\mathbf{k})}{\gamma_u}[\mathbb{I}_{mn}^{(u^<)} - H\mathbb{I}_{mn}^{(b^<)} - \langle\mathbb{I}_{mn}^{(u^<)}\rangle + H\langle\mathbb{I}_{mn}^{(b^<)}\rangle]$$
$$- HRo\left(Ro\tilde{\beta}\right)\frac{\mathbf{k}\cdot\langle\mathbf{B}\rangle}{\gamma_u\gamma_\eta}k_j\left[\mathbb{I}_{ij}^{(u^<b^<)} - \left\langle\mathbb{I}_{ij}^{(u^<b^<)}\right\rangle\right] \tag{4.51a}$$
$$- \mathrm{i}\frac{Ro}{\gamma_u}P_{imn}[\mathbb{J}_{mn}^{(u)} - H\mathbb{J}_{mn}^{(b)}] - HRo\left(Ro\tilde{\beta}\right)\frac{\mathbf{k}\cdot\langle\mathbf{B}\rangle}{\gamma_u\gamma_\eta}k_j\mathbb{J}_{ij}^{(ub)} + R_i^{(u)},$$

$$\hat{b}_i^>(\mathbf{q}) = \mathrm{i}Ro\tilde{\beta}\frac{\mathbf{k}\cdot\langle\mathbf{B}\rangle}{\gamma_u\gamma_\eta}\hat{f}_i^>(\mathbf{q}) + \mathrm{i}\frac{Ro}{\gamma_\eta}k_j\left[\mathbb{I}_{ij}^{(u^<b^<)} - \left\langle\mathbb{I}_{ij}^{(u^<b^<)}\right\rangle\right]$$
$$+ \frac{1}{2}Ro\left(Ro\tilde{\beta}\right)\frac{\mathbf{k}\cdot\langle\mathbf{B}\rangle}{\gamma_u\gamma_\eta}P_{imn}(\mathbf{k})[\mathbb{I}_{mn}^{(u^<)} - H\mathbb{I}_{mn}^{(b^<)} - \langle\mathbb{I}_{mn}^{(u^<)}\rangle + H\langle\mathbb{I}_{mn}^{(b^<)}\rangle] \tag{4.51b}$$
$$+ \mathrm{i}\frac{Ro}{\gamma_\eta}k_j\mathbb{J}_{ij}^{(ub)} + Ro\left(Ro\tilde{\beta}\right)\frac{\mathbf{k}\cdot\langle\mathbf{B}\rangle}{\gamma_u\gamma_\eta}P_{imn}[\mathbb{J}_{mn}^{(u)} - H\mathbb{J}_{mn}^{(b)}] + R_i^{(b)},$$

where $\langle\cdot\rangle_c$ denotes conditional average over the first shell ($k_1 < k < K_{\max}$) statistical subensemble, described at the beginning of the next section (for details see [9–11]); furthermore, we have defined

$$\mathbb{J}_{mn}^{(u)}(\mathbf{q}) = \theta_{K_\nu}\int_{K_e}^{K_\nu} d^4q'\hat{u}_m^<(\mathbf{q}')\hat{u}_n^>(\mathbf{q}-\mathbf{q}'), \tag{4.52a}$$

$$\mathbb{J}_{mn}^{(b)}(\mathbf{q}) = \theta_{K_\eta}\int_{K_e}^{K_\eta} d^4q'\hat{b}_m^<(\mathbf{q}')\hat{b}_n^>(\mathbf{q}-\mathbf{q}'), \tag{4.52b}$$

$$\mathbb{J}_{ij}^{(ub)}(\mathbf{q}) = \epsilon_{ijk}\epsilon_{kmn}\theta_K\int_{K_e}^{K} d^4q'\left[\hat{u}_m^<(\mathbf{q}')\hat{b}_n^>(\mathbf{q}-\mathbf{q}') + \hat{u}_m^>(\mathbf{q}')\hat{b}_n^<(\mathbf{q}-\mathbf{q}')\right], \tag{4.52c}$$

and the reminders in (4.51a and b) are given by

$$R_i^{(u)} = -\frac{Ro}{\gamma_u}\left\{\frac{1}{2}\mathrm{i}P_{imn}[\mathbb{I}_{mn}^{(u^>)} - H\mathbb{I}_{mn}^{(b^>)}] + HRo\tilde{\beta}\frac{\mathbf{k}\cdot\langle\mathbf{B}\rangle}{\gamma_\eta}k_j\mathbb{I}_{ij}^{(u^>b^>)}\right\}$$
$$+ \frac{Ro}{\gamma_u}\left\{\frac{1}{2}\mathrm{i}P_{imn}[\langle\mathbb{I}_{mn}^{(u^>)}\rangle - H\langle\mathbb{I}_{mn}^{(b^>)}\rangle] + HRo\tilde{\beta}\frac{\mathbf{k}\cdot\langle\mathbf{B}\rangle}{\gamma_\eta}k_j\left\langle\mathbb{I}_{ij}^{(u^>b^>)}\right\rangle\right\}, \tag{4.53a}$$

$$
\begin{aligned}
R_i^{(b)} = {} & \mathrm{i}\frac{Ro}{\gamma_\eta}k_j\left[\mathbb{I}_{ij}^{(u^>b^>)} - \left\langle \mathbb{I}_{ij}^{(u^>b^>)} \right\rangle\right] \\
& + \frac{1}{2}Ro\left(Ro\tilde{\beta}\right)\frac{\mathbf{k}\cdot\langle\mathbf{B}\rangle}{\gamma_u\gamma_\eta}P_{imn}[\mathbb{I}_{mn}^{(u^>)} - H\mathbb{I}_{mn}^{(b^>)} - \langle\mathbb{I}_{mn}^{(u^>)}\rangle + H\langle\mathbb{I}_{mn}^{(b^>)}\rangle] \\
& - \mathrm{i}HRo\left(Ro\tilde{\beta}\right)^2\frac{(\mathbf{k}\cdot\langle\mathbf{B}\rangle)^2}{\gamma_u\gamma_\eta^2}k_j\left[\mathbb{I}_{ij}^{(u^<b^<)} + \mathbb{J}_{ij}^{(ub)} + \mathbb{I}_{ij}^{(u^>b^>)}\right. \\
& \qquad \left. -\left\langle \mathbb{I}_{ij}^{(u^<b^<)} \right\rangle - \left\langle \mathbb{I}_{ij}^{(u^>b^>)} \right\rangle\right].
\end{aligned}
\tag{4.53b}
$$

The remainders will be neglected either on the basis of generating only triple-order statistical correlations, as in the case of all the terms that are second order in $\mathbf{u}^>$ or $\mathbf{b}^>$, or because of the kept order of accuracy in the asymptotic limit $Ro \ll Ro\tilde{\beta} \ll 1$, which allows us to neglect terms of the order $\mathcal{O}(Ro(Ro\tilde{\beta})^2)$.

In the next section we apply the weakly nonlinear renormalization method, which consists of the following steps. First we introduce (4.51a and b) into (4.50a and b) and calculate the dynamical effect of short-wavelength components $\hat{u}_i^>(\mathbf{k}, \omega)$ and $\hat{b}_i^>(\mathbf{k}, \omega)$ on the evolution of $\hat{u}_i^<(\mathbf{k}, \omega)$ and $\hat{b}_i^<(\mathbf{k}, \omega)$ (long-wavelength modes). This results in corrections to some of the terms in equations (4.50a and b), but also generates terms with new structure. Therefore, a next step is necessary, involving calculation of the effect of the next shell $k_2 = k_1 - \delta k < k < k_1$ (new short-wavelength modes) on the modes with $k < k_1 - \delta k$ (new long-wavelength modes). We can then take the limit of infinitesimally narrow wave number bands $\delta k \to 0$, which leads to differential recursion relations for all the coupling constants introduced into the equations for long-wavelength modes by couplings of the short-wavelength ones. Moffatt (1983) [7] obtained such equations but in the kinematic case with the turbulence flow given beforehand, unaffected by the Lorentz force (see also [8], where in the non-magnetic case the leading order correction from short-wavelength modes in the Navier–Stokes equation was calculated and shown to be proportional to $k^2\hat{u}_i^<\delta k$ thus creating viscosity correction; the turbulent viscosity was then obtained from an equation of the form $\mathrm{d}\nu_{\mathrm{turb}}/\mathrm{d}k = f(k)$ with an 'initial' condition $\nu_{\mathrm{turb}}(k = K_\nu) = \nu$). In the case at hand under the assumptions $Ro \ll Ro\tilde{\beta} \ll 1$, the explicit calculation of two initial steps of the renormalization procedure is enough to derive the final differential recursion relations with satisfactory accuracy. The details of the procedure are provided in the following section.

4.3.1 The recursion differential equations

The details of the renormalization procedure applied in order to obtain a closed system of the mean field equations are given here. First of all, we clarify how the ensemble averaging should be understood and explain the concept of a conditional average over a statistical subensemble for short-wavelength modes. We adopt the approach of [11] (cf also [9, 10]). The essential idea of this approach is the introduction of a subensemble of flow realizations including near-chaotic statistical properties for the short-wavelength shell $k_n < k \leqslant k_{n-1}$, but remaining quasi-deterministic for $k \leqslant k_n$. The subensemble average can be precisely defined and then, utilizing the assumption, that in the turbulent cascade the energy transfer in the Fourier space is local, the following crucial properties can be proved

$$\langle \hat{\mathbf{u}}^{<}(\mathbf{q})\rangle_c = \hat{\mathbf{u}}^{<}(\mathbf{q}), \qquad \langle \hat{\mathbf{u}}^{<}(\mathbf{q})\hat{\mathbf{u}}^{<}(\mathbf{q}')\rangle_c \approx \hat{\mathbf{u}}^{<}(\mathbf{q})\hat{\mathbf{u}}^{<}(\mathbf{q}'), \tag{4.54a}$$

$$\langle \hat{\mathbf{u}}^{>}(\mathbf{q}')\rangle_c \approx \langle \hat{\mathbf{u}}^{>}(\mathbf{q}')\rangle = 0, \qquad \langle \hat{\mathbf{u}}^{<}(\mathbf{q})\hat{\mathbf{u}}^{>}(\mathbf{q}')\rangle_c \approx \hat{\mathbf{u}}^{<}(\mathbf{q})\langle \hat{\mathbf{u}}^{>}(\mathbf{q}')\rangle_c \approx 0, \tag{4.54b}$$

$$\langle \hat{\mathbf{u}}^{>}(\mathbf{q})\hat{\mathbf{u}}^{>}(\mathbf{q}')\rangle_c \approx \langle \hat{\mathbf{u}}^{>}(\mathbf{q})\hat{\mathbf{u}}^{>}(\mathbf{q}')\rangle. \tag{4.54c}$$

For details see particularly section IV and the beginning of section V in [11].

We now substitute the expressions for short-wavelength modes from (4.51a and b) into the conditional averages in the equations for long-wavelength modes in (4.50a and b). Neglecting higher order correlations of the type $\left\langle \hat{u}_i^{>}\hat{u}_j^{>}\hat{f}_k^{>}\right\rangle_c$ etc (which together with neglection of $\mathcal{O}(Ro(Ro\tilde{\beta})^2)$ terms eliminates the remainders in (4.51a and b)), and using $\left\langle \hat{f}_i^{>}\right\rangle_c = 0$, $\langle \hat{u}_i^{<}\rangle_c = \hat{u}_i^{<}$ and $\left\langle \hat{b}_i^{<}\right\rangle_c = \hat{b}_i^{<}$ one obtains

$$\begin{aligned}
&\langle \hat{u}_m^{>}(\mathbf{q}')\hat{u}_n^{>}(\mathbf{q}-\mathbf{q}')\rangle_c - \langle \hat{u}_m^{>}(\mathbf{q}')\hat{u}_n^{>}(\mathbf{q}-\mathbf{q}')\rangle = \\
&- \frac{\mathrm{i}Ro}{\gamma_u(\mathbf{q}')\gamma_u(\mathbf{q}-\mathbf{q}')} P_{nqp}(\mathbf{k}-\mathbf{k}')\left[\left\langle \hat{f}_m^{>}(\mathbf{q}')\mathbb{J}_{qp}^{(u)}(\mathbf{q}-\mathbf{q}')\right\rangle_c - H\left\langle \hat{f}_m^{>}(\mathbf{q}')\mathbb{J}_{qp}^{(b)}(\mathbf{q}-\mathbf{q}')\right\rangle_c\right] \\
&- \frac{\mathrm{i}Ro}{\gamma_u(\mathbf{q}')\gamma_u(\mathbf{q}-\mathbf{q}')} P_{mqp}(\mathbf{k}')\left[\left\langle \hat{f}_n^{>}(\mathbf{q}-\mathbf{q}')\mathbb{J}_{qp}^{(u)}(\mathbf{q}')\right\rangle_c - H\left\langle \hat{f}_n^{>}(\mathbf{q}-\mathbf{q}')\mathbb{J}_{qp}^{(b)}(\mathbf{q}')\right\rangle_c\right] \\
&- HRo\left(Ro\tilde{\beta}\right)\frac{(\mathbf{k}-\mathbf{k}')\cdot\langle\mathbf{B}\rangle}{\gamma_u(\mathbf{q}')\gamma_u(\mathbf{q}-\mathbf{q}')\gamma_\eta(\mathbf{q}-\mathbf{q}')}\left(k_j-k_j'\right)\left\langle \hat{f}_m^{>}(\mathbf{q}')\mathbb{J}_{nj}^{(ub)}(\mathbf{q}-\mathbf{q}')\right\rangle_c \\
&- HRo\left(Ro\tilde{\beta}\right)\frac{\mathbf{k}'\cdot\langle\mathbf{B}\rangle}{\gamma_u(\mathbf{q}-\mathbf{q}')\gamma_u(\mathbf{q}')\gamma_\eta(\mathbf{q}')} k_j'\left\langle \hat{f}_n^{>}(\mathbf{q}-\mathbf{q}')\mathbb{J}_{mj}^{(ub)}(\mathbf{q}')\right\rangle_c + \mathcal{O}(Ro^2),
\end{aligned} \tag{4.55a}$$

$$\left\langle \hat{u}_i^{>}(\mathbf{q}')\hat{u}_j^{>}(\mathbf{q}-\mathbf{q}')\right\rangle = \frac{1}{\gamma_u(\mathbf{q}')\gamma_u(\mathbf{q}-\mathbf{q}')}\left\langle \hat{f}_i^{>}(\mathbf{q}')\hat{f}_j^{>}(\mathbf{q}-\mathbf{q}')\right\rangle, \tag{4.55b}$$

$$\begin{aligned}
&\left\langle \hat{b}_m^{>}(\mathbf{q}')\hat{b}_n^{>}(\mathbf{q}-\mathbf{q}')\right\rangle_c - \left\langle \hat{b}_m^{>}(\mathbf{q}')\hat{b}_n^{>}(\mathbf{q}-\mathbf{q}')\right\rangle = \\
&- \frac{Ro\left(Ro\tilde{\beta}\right)(\mathbf{k}'\cdot\langle\mathbf{B}\rangle)}{\gamma_u(\mathbf{q}')\gamma_\eta(\mathbf{q}')\gamma_\eta(\mathbf{q}-\mathbf{q}')}\left(k_j-k_j'\right)\left\langle \hat{f}_m^{>}(\mathbf{q}')\mathbb{J}_{nj}^{(ub)}(\mathbf{q}-\mathbf{q}')\right\rangle_c \\
&- \frac{Ro\left(Ro\tilde{\beta}\right)((\mathbf{k}-\mathbf{k}')\cdot\langle\mathbf{B}\rangle)}{\gamma_u(\mathbf{q}-\mathbf{q}')\gamma_\eta(\mathbf{q}-\mathbf{q}')\gamma_\eta(\mathbf{q}')} k_j'\left\langle \hat{f}_n^{>}(\mathbf{q}-\mathbf{q}')\mathbb{J}_{mj}^{(ub)}(\mathbf{q}')\right\rangle_c \\
&+ \mathcal{O}\left(Ro^2, Ro\left(Ro\tilde{\beta}\right)^2\right),
\end{aligned} \tag{4.55c}$$

$$\left\langle \hat{b}_i^{>}(\mathbf{q}')\hat{b}_j^{>}(\mathbf{q}-\mathbf{q}')\right\rangle$$
$$= -\left(Ro\tilde{\beta}\right)^2 \frac{(\mathbf{k}'\cdot\langle\mathbf{B}\rangle)((\mathbf{k}-\mathbf{k}')\cdot\langle\mathbf{B}\rangle)}{\gamma_u(\mathbf{q}')\gamma_\eta(\mathbf{q}')\gamma_u(\mathbf{q}-\mathbf{q}')\gamma_\eta(\mathbf{q}-\mathbf{q}')}\left\langle \hat{f}_i^{>}(\mathbf{q}')\hat{f}_j^{>}(\mathbf{q}-\mathbf{q}')\right\rangle, \tag{4.55d}$$

$$\begin{aligned}
&\left\langle \hat{u}_m^{>}(\mathbf{q}')\hat{b}_n^{>}(\mathbf{q}-\mathbf{q}')\right\rangle_c - \left\langle \hat{u}_m^{>}(\mathbf{q}')\hat{b}_n^{>}(\mathbf{q}-\mathbf{q}')\right\rangle = \\
&\frac{\mathrm{i}Ro}{\gamma_u(\mathbf{q}')\gamma_\eta(\mathbf{q}-\mathbf{q}')}(k_s - k_s')\left\langle \hat{f}_m^{>}(\mathbf{q}')\mathbb{J}_{ns}^{(ub)}(\mathbf{q}-\mathbf{q}')\right\rangle_c \\
&+\frac{Ro\left(Ro\tilde{\beta}\right)(\mathbf{k}-\mathbf{k}')\cdot\langle\mathbf{B}\rangle}{\gamma_u(\mathbf{q}')\gamma_u(\mathbf{q}-\mathbf{q}')\gamma_\eta(\mathbf{q}-\mathbf{q}')}P_{nqp}(\mathbf{k}-\mathbf{k}') \\
&\times\left[\left\langle \hat{f}_m^{>}(\mathbf{q}')\mathbb{J}_{qp}^{(u)}(\mathbf{q}-\mathbf{q}')\right\rangle_c - H\left\langle \hat{f}_m^{>}(\mathbf{q}')\mathbb{J}_{qp}^{(b)}(\mathbf{q}-\mathbf{q}')\right\rangle_c\right] \\
&+\frac{Ro\left(Ro\tilde{\beta}\right)(\mathbf{k}-\mathbf{k}')\cdot\langle\mathbf{B}\rangle}{\gamma_u(\mathbf{q}')\gamma_u(\mathbf{q}-\mathbf{q}')\gamma_\eta(\mathbf{q}-\mathbf{q}')}P_{mqp}(\mathbf{k}') \\
&\times\left[\left\langle \hat{f}_n^{>}(\mathbf{q}-\mathbf{q}')\mathbb{J}_{qp}^{(u)}(\mathbf{k}')\right\rangle_c - H\left\langle \hat{f}_n^{>}(\mathbf{q}-\mathbf{q}')\mathbb{J}_{qp}^{(b)}(\mathbf{k}')\right\rangle_c\right] \\
&+\mathcal{O}\left(Ro^2, Ro\left(Ro\tilde{\beta}\right)^2\right),
\end{aligned} \tag{4.55e}$$

$$\epsilon_{kmn}\left\langle \hat{u}_m^{>}(\mathbf{q}')\hat{b}_n^{>}(\mathbf{q}-\mathbf{q}')\right\rangle = \frac{\mathrm{i}Ro\tilde{\beta}\epsilon_{kmn}(\mathbf{k}-\mathbf{k}')\cdot\langle\mathbf{B}\rangle}{\gamma_u(\mathbf{q}')\gamma_u(\mathbf{q}-\mathbf{q}')\gamma_\eta(\mathbf{q}-\mathbf{q}')}\left\langle \hat{f}_m^{>}(\mathbf{q}')\hat{f}_n^{>}(\mathbf{q}-\mathbf{q}')\right\rangle. \tag{4.55f}$$

The term (4.55c) accounts for the Lorentz force effect of short-wavelength fluctuations on the long-wavelength ones (fluctuational Lorentz force acting on fluctuations), whereas the terms (4.55d) and (4.55b) contribute directly to the effective mean Lorentz force, which acts on the mean field $\langle\mathbf{U}\rangle$ (mean fluctuational Lorentz force acting on the mean velocity); naturally, in order to get the Lorentz force from (4.55b) one needs to expand the γ_u^{-1} factors in powers of $(Ro\tilde{\beta})^2$. The former, that is, the fluctuational Lorentz force acting on the fluctuational velocity, is of less importance, as it turns out to influence the α-effect and the effective mean Lorentz force only at higher, negligible orders. It is therefore desirable for clarity of the cumbersome calculations to narrow down the accuracy and neglect all the terms of the order $\mathcal{O}(Ro(Ro\tilde{\beta}))$. This will still allow us to obtain the leading order form of the turbulent EMF and the mean Lorentz force under the assumption $Ro \ll Ro\tilde{\beta} \ll 1$ and *with inclusion of the nonlinear evolution of turbulent fluctuations.* Substituting once again for $\hat{\mathbf{u}}^{>}$ and $\hat{\mathbf{b}}^{>}$ from (4.51a and b) into all the $\mathbb{J}$-terms in (4.55a) and (4.55e) and making use of the symmetry $\mathbf{q}' \mapsto \mathbf{q}-\mathbf{q}'$ under the integral $\int d^4q'$ one obtains

$$\theta_{K_\nu}\int \mathrm{d}^4q'\left[\langle \hat{u}^>_m(\mathbf{q}')\hat{u}^>_n(\mathbf{q}-\mathbf{q}')\rangle_c - \langle \hat{u}^>_m(\mathbf{q}')\hat{u}^>_n(\mathbf{q}-\mathbf{q}')\rangle\right] =$$

$$\begin{aligned} &-\mathrm{i}Ro\theta_{K_\nu}\int \mathrm{d}^4q'\int \mathrm{d}^4q''\frac{\hat{u}^<_q(\mathbf{q}'')P_{nqp}(\mathbf{k}-\mathbf{k}')\left\langle\hat{f}^>_m(\mathbf{q}')\hat{f}^>_p(\mathbf{q}-\mathbf{q}'-\mathbf{q}'')\right\rangle_c}{\gamma_u(\mathbf{q}')\gamma_u(\mathbf{q}-\mathbf{q}')\gamma_u(\mathbf{q}-\mathbf{q}'-\mathbf{q}'')} \\ &-\mathrm{i}Ro\theta_{K_\nu}\int \mathrm{d}^4q'\int^{\Lambda_\nu} \mathrm{d}^4q''\frac{\hat{u}^<_q(\mathbf{q}'')P_{mqp}(\mathbf{k}-\mathbf{k}')\left\langle\hat{f}^>_n(\mathbf{q}')\hat{f}^>_p(\mathbf{q}-\mathbf{q}'-\mathbf{q}'')\right\rangle_c}{\gamma_u(\mathbf{q}')\gamma_u(\mathbf{q}-\mathbf{q}')\gamma_u(\mathbf{q}-\mathbf{q}'-\mathbf{q}'')} \\ &+\mathcal{O}\left(Ro(Ro\tilde{\beta})\right), \end{aligned} \tag{4.56a}$$

$$\theta_{K_\eta}\int \mathrm{d}^4q'\left[\left\langle \hat{b}^>_m(\mathbf{q}')\hat{b}^>_n(\mathbf{q}-\mathbf{q}')\right\rangle_c - \left\langle \hat{b}^>_m(\mathbf{q}')\hat{b}^>_n(\mathbf{q}-\mathbf{q}')\right\rangle\right] = \mathcal{O}\left(Ro\left(Ro\tilde{\beta}\right)\right), \tag{4.56b}$$

$$\theta_K\int \mathrm{d}^4q'\left[\left\langle \hat{u}^>_m(\mathbf{q}')\hat{b}^>_n(\mathbf{q}-\mathbf{q}')\right\rangle_c - \left\langle \hat{u}^>_m(\mathbf{q}')\hat{b}^>_n(\mathbf{q}-\mathbf{q}')\right\rangle\right] =$$

$$\begin{aligned} &\mathrm{i}Ro\theta_K\epsilon_{nsr}\epsilon_{rqp}\int \mathrm{d}^4q'\int \mathrm{d}^4q''\frac{\hat{b}^<_p(\mathbf{q}-\mathbf{q}'-\mathbf{q}'')\left\langle\hat{f}^>_m(\mathbf{q}')\hat{f}^>_q(\mathbf{q}'')\right\rangle_c}{\gamma_u(\mathbf{q}')\gamma_u(\mathbf{q}'')\gamma_\eta(\mathbf{q}-\mathbf{q}')}(k_s-k_s') \\ &+\mathcal{O}\left(Ro\left(Ro\tilde{\beta}\right)\right), \end{aligned} \tag{4.56c}$$

where now the q'-integrals in (4.56a) are taken over an intersection of the domains $k_1 < k' < K_{\max}$ and $k_1 < |\mathbf{k}-\mathbf{k}'| < K_{\max}$. Following the approach of [8] and [15] we calculate the q'-integrals to lowest non-trivial order in the distant-interaction limit

$$\frac{k}{k'} \to 0, \quad \frac{\omega}{\omega'} \to 0, \tag{4.57}$$

which stems from the assumption of local energy transfer in the Fourier spectrum of a turbulent cascade. However, contrary to the purely hydrodynamic case (without the magnetic field) considered in the aforementioned works, in the magnetic case the lowest non-trivial order in k/k' and ω/ω' is obtained by simply setting $\mathbf{k}=0$ and $\omega=0$ in the integrands[1]. At this lowest non-trivial order the domain of integration simplifies to

$$\{\mathbf{k}:\ k_1 < q' < K_{\max}\}, \tag{4.58}$$

and higher orders involve inclusion of contributions from the non-trivial shape of the integration domain, which will not be considered here. The total renormalized

[1] In the non-magnetic case considered by [8] the integrals vanish at the order $(k/k')^0$ and expansion in powers of k/k' of the integrands to the first order is necessary; this results in a viscous correction of the order $\sim\nu k^2$.

corrections from short-wavelength modes in (4.50a and b) are linear in the wave vector $\mathbf{k}$, which implies that in the MHD case the lowest non-trivial order in distant interactions does not produce any corrections to diffusivities. Turbulent viscosity and magnetic diffusivity could only be obtained at higher orders (cf [18]). It follows that, at the kept order of accuracy, we retain the molecular viscosity and magnetic diffusivity everywhere, i.e. also in the mean field equations.

We also utilize the assumption $Ro\tilde{\beta} \ll 1$ to expand the inverse γ_u-factors

$$\frac{1}{\gamma_u(\omega', \mathbf{k}')\gamma_u(\omega - \omega', \mathbf{k}')} = \frac{1}{\gamma_\nu(\omega', k')\gamma_\nu(\omega - \omega', k')} - H\left(Ro\tilde{\beta}\right)^2 (\mathbf{k}' \cdot \langle\mathbf{B}\rangle)^2 \left[\frac{1}{\gamma_\nu(\omega', k')\gamma_\nu(\omega - \omega', k')^2\gamma_\eta(\omega - \omega', k')} + \frac{1}{\gamma_\nu(\omega', k')^2\gamma_\nu(\omega - \omega', k')\gamma_\eta(\omega', k')}\right] + \mathcal{O}\left(\left(Ro\tilde{\beta}\right)^4\right), \tag{4.59}$$

where

$$\gamma_u = -\mathrm{i}\omega + E_\nu k^2 + H\left(Ro\tilde{\beta}\right)^2 \frac{(\mathbf{k} \cdot \langle\mathbf{B}\rangle)^2}{\gamma_\eta}, \tag{4.60}$$

$$\gamma_\nu(\omega, k) = -\mathrm{i}\omega + E_\nu k^2, \quad \gamma_\eta(\omega, k) = -\mathrm{i}\omega + E_\eta k^2, \tag{4.61}$$

and we recall that the mean velocity term, $\mathbf{k} \cdot \langle\mathbf{U}\rangle$, is absorbed into the frequency. Substituting for the force correlations from (4.45) into (4.55b), (4.55d) and (4.55f), taking the first of the limits in (4.57) describing distant wave-number interactions $k/k' \to \infty$, and making use of (4.59) and

$$P_{imn}(\mathbf{k})\delta_{mn} = 0, \quad P_{imn}(\mathbf{k})\epsilon_{mnk} = 0, \quad \text{for all } i,\ k, \tag{4.62a}$$

$$\epsilon_{ijk}\epsilon_{kmn} = \delta_{im}\delta_{jn} - \delta_{in}\delta_{jm}, \tag{4.62b}$$

$$\int \mathrm{d}\mathring{\Omega} \underbrace{k_m \dots k_n k_k}_{N} = 0, \quad \text{for any odd } N \text{ and all } m, \dots, n, k, \tag{4.62c}$$

$$\int k^2 \mathrm{d}\mathring{\Omega} \frac{k_m k_n}{k^4} = \frac{4\pi}{3}\delta_{mn}, \tag{4.62d}$$

$$\int k^2 \mathrm{d}\mathring{\Omega}\frac{k_m k_n k_p k_q}{k^6} = \frac{4\pi}{15}\left(\delta_{mn}\delta_{pq} + \delta_{mp}\delta_{nq} + \delta_{mq}\delta_{np}\right), \tag{4.62e}$$

where $\mathring{\Omega}$ denotes a solid angle, one obtains

$$\begin{aligned}
&- Ro\theta_{K_\nu}\int \mathrm{d}^4q'\left\langle \hat{u}_i^{>}(\mathbf{q}')\hat{u}_j^{>}(\mathbf{q}-\mathbf{q}')\right\rangle \\
&= \delta(\mathbf{k})\delta(\omega)\theta_{K_\nu}\int \frac{\mathrm{d}^4q'}{k'^3}k'_q k'_p P_{ij}(\mathbf{k}')\left[\frac{Ro\left(Ro\tilde{\beta}\right)^2 HD_0\langle B\rangle_q\langle B\rangle_p}{\gamma_\nu(\omega', k')\gamma_\nu(\omega-\omega', k')^2\gamma_\eta(\omega-\omega', k')}\right. \\
&\qquad\left. + \frac{Ro\left(Ro\tilde{\beta}\right)^2 HD_0\langle B\rangle_q\langle B\rangle_p}{\gamma_\nu(\omega', k')^2\gamma_\nu(\omega-\omega', k')\gamma_\eta(\omega', k')}\right] \\
&+ \delta(\mathbf{k})[\delta(\omega+\tilde{\omega}) - \delta(\omega-\tilde{\omega})]\frac{\theta_{K_\nu}}{2\mathrm{i}}\int \frac{\mathrm{d}^4q'}{k'^3}k'_q k'_p P_{ij}(\mathbf{k}')\times \\
&\left[\frac{Ro\left(Ro\tilde{\beta}\right)^2 H\Gamma D_0\langle B\rangle_q\langle B\rangle_p}{\gamma_\nu(\omega', k')\gamma_\nu(\omega-\omega', k')^2\gamma_\eta(\omega-\omega', k')}\right. \\
&\qquad\left. + \frac{Ro\left(Ro\tilde{\beta}\right)^2 H\Gamma D_0\langle B\rangle_q\langle B\rangle_p}{\gamma_\nu(\omega', k')^2\gamma_\nu(\omega-\omega', k')\gamma_\eta(\omega', k')}\right] + \mathcal{O}\left(Ro\left(Ro\tilde{\beta}\right)^4\right),
\end{aligned} \tag{4.63a}$$

$$\begin{aligned}
&RoH\theta_{K_\eta}\int \mathrm{d}^4q'\left\langle \hat{b}_i^{>}(\mathbf{q}')\hat{b}_j^{>}(\mathbf{q}-\mathbf{q}')\right\rangle \\
&= \delta(\mathbf{k})\delta(\omega)\theta_{K_\eta}\int \frac{\mathrm{d}^4q'}{k'^3}\frac{Ro\left(Ro\tilde{\beta}\right)^2 HD_0\langle B\rangle_q\langle B\rangle_p k'_q k'_p P_{ij}(\mathbf{k}')}{\gamma_\nu(\omega', k')\gamma_\eta(\omega', k')\gamma_\nu(\omega-\omega', k')\gamma_\eta(\omega-\omega', k')} \\
&+ \delta(\mathbf{k})[\delta(\omega+\tilde{\omega}) - \delta(\omega-\tilde{\omega})]\frac{\theta_{K_\eta}}{2\mathrm{i}}Ro\left(Ro\tilde{\beta}\right)^2 H\Gamma D_0\langle B\rangle_q\langle B\rangle_p\times \\
&\qquad\int \frac{\mathrm{d}^4q'}{k'^3}\frac{k'_q k'_p P_{ij}(\mathbf{k}')}{\gamma_\nu(\omega', k')\gamma_\eta(\omega', k')\gamma_\nu(\omega-\omega', k')\gamma_\eta(\omega-\omega', k')} \\
&+ \mathcal{O}\left(Ro\left(Ro\tilde{\beta}\right)^4\right),
\end{aligned} \tag{4.63b}$$

$$
\begin{aligned}
&Ro\tilde{\beta}^{-1}\epsilon_{kmn}\theta_K \int \mathrm{d}^4q' \left\langle \hat{u}_m^{>}(\mathbf{q}')\hat{b}_n^{>}(\mathbf{q}-\mathbf{q}') \right\rangle = \\
&\delta(\mathbf{k})\delta(\omega)2\theta_K \int \frac{\mathrm{d}^4q'}{k'^5} \frac{Ro^2 D_1 \langle B\rangle_p k'_p k'_k}{\gamma_\nu(\omega', k')\gamma_\nu(\omega-\omega', k')\gamma_\eta(\omega-\omega', k')} \\
&- \delta(\mathbf{k})(\delta(\omega+\tilde{\omega}) - \delta(\omega-\tilde{\omega}))\mathrm{i}\theta_K \int \frac{\mathrm{d}^4q'}{k'^5} \frac{Ro^2 \Gamma D_1 \langle B\rangle_p k'_p k'_k}{\gamma_\nu(\omega', k')\gamma_\nu(\omega-\omega', k')\gamma_\eta(\omega-\omega', k')} \\
&- 2\delta(\mathbf{k})\delta(\omega)\theta_K \int \frac{\mathrm{d}^4q'}{k'^5} \frac{k'_p k'_q k'_r k'_k}{\gamma_\eta(\omega-\omega', k')} \left[\frac{Ro^2\left(Ro\tilde{\beta}\right)^2 H D_1 \langle B\rangle_p \langle B\rangle_q \langle B\rangle_r}{\gamma_\nu(\omega', k')\gamma_\nu(\omega-\omega', k')^2\gamma_\eta(\omega-\omega', k')} \right. \\
&\qquad \left. + \frac{Ro^2\left(Ro\tilde{\beta}\right)^2 H D_1 \langle B\rangle_p \langle B\rangle_q \langle B\rangle_r}{\gamma_\nu(\omega', k')^2\gamma_\nu(\omega-\omega', k')\gamma_\eta(\omega', k')} \right] \\
&+ \delta(\mathbf{k})[\delta(\omega+\tilde{\omega}) - \delta(\omega-\tilde{\omega})]\theta_K \int^{\Lambda} \frac{\mathrm{d}^4q'}{k'^5} \frac{\mathrm{i}Ro^2\left(Ro\tilde{\beta}\right)^2 H\Gamma D_1 k'_p k'_q k'_r k'_k}{\gamma_\eta(\omega-\omega', k')\gamma_\nu(\omega-\omega', k')} \times \\
&\left[\frac{\langle B\rangle_p \langle B\rangle_q \langle B\rangle_r}{\gamma_\nu(\omega', k')\gamma_\nu(\omega-\omega', k')\gamma_\eta(\omega-\omega', k')} + \frac{\langle B\rangle_p \langle B\rangle_q \langle B\rangle_r}{\gamma_\nu^2(\omega', k')\gamma_\eta(\omega', k')} \right] \\
&+ \mathcal{O}\left(Ro^2\left(Ro\tilde{\beta}\right)^4\right).
\end{aligned} \tag{4.63c}
$$

Since our main goal is obtaining the expression for the mean EMF in strong turbulence, the effective mean Lorentz force does not need to be calculated with high accuracy. It follows, that the contributions to the mean Lorentz force from the nonlinear interactions of fluctuations in (4.63a and b) are provided only to the first non-vanishing order (the next order, $\mathcal{O}(Ro(Ro\tilde{\beta})^4)$, can be obtained in a straightforward manner by the use of (4.59)).

Next, from (4.62c) it is obvious, that the leading order term in (4.56a) vanishes. It follows that the corrections from short-wavelength modes in the fluctuational equations (4.56a–c) take the form

$$
\begin{aligned}
&-\frac{\mathrm{i}}{2}Ro\theta_{K_\nu}P_{imn}(\mathbf{k}) \int \mathrm{d}^4q' \left[\langle \hat{u}_m^{>}(\mathbf{q}')\hat{u}_n^{>}(\mathbf{q}-\mathbf{q}')\rangle_c - \langle \hat{u}_m^{>}(\mathbf{q}')\hat{u}_n^{>}(\mathbf{q}-\mathbf{q}')\rangle \right] \\
&\qquad = \mathcal{O}\left(Ro^2\left(Ro\tilde{\beta}\right)\right),
\end{aligned} \tag{4.64a}
$$

$$
\begin{aligned}
&\frac{\mathrm{i}}{2}RoH\theta_{K_\eta}P_{imn}(\mathbf{k}) \int \mathrm{d}^4q' \left[\left\langle \hat{b}_m^{>}(\mathbf{q}')\hat{b}_n^{>}(\mathbf{q}-\mathbf{q}') \right\rangle_c - \left\langle \hat{b}_m^{>}(\mathbf{q}')\hat{b}_n^{>}(\mathbf{q}-\mathbf{q}') \right\rangle \right] \\
&\qquad = \mathcal{O}\left(Ro^2\left(Ro\tilde{\beta}\right)\right),
\end{aligned} \tag{4.64b}
$$

$$
\begin{aligned}
&\mathrm{i}Ro\theta_K\epsilon_{ijk}\epsilon_{kmn}k_j\int \mathrm{d}^4q'\Big[\Big\langle \hat{u}^{>}_m(\mathbf{q}')\hat{b}^{>}_n(\mathbf{q}-\mathbf{q}')\Big\rangle_c - \Big\langle \hat{u}^{>}_m(\mathbf{q}')\hat{b}^{>}_n(\mathbf{q}-\mathbf{q}')\Big\rangle\Big]\\
&= 2\mathrm{i}\theta_K k_j\epsilon_{ijk}\int \frac{\mathrm{d}^4q'}{k'^5}\frac{Ro^2D_1\hat{b}^{<}_p(\omega,\mathbf{k})k'_kk'_p}{\gamma_\nu(\omega',k')\gamma_\nu(-\omega',k')\gamma_\eta(\omega-\omega',k')}\\
&+\theta_K k_j\epsilon_{ijk}\int \frac{\mathrm{d}^4q'}{k'^5}\frac{Ro^2\Gamma D_1k'_kk'_p}{\gamma_\nu(\omega',k')\gamma_\eta(\omega-\omega',k')}\left[\frac{\hat{b}^{<}_p(\omega+\tilde{\omega},\mathbf{k})}{\gamma_\nu(-\omega'-\tilde{\omega},k')}\right.\\
&\qquad\left.-\frac{\hat{b}^{<}_p(\omega-\tilde{\omega},\mathbf{k})}{\gamma_\nu(-\omega'+\tilde{\omega},k')}\right]+\mathcal{O}\big(Ro^2\big(Ro\tilde{\beta}\big)\big).
\end{aligned}
\tag{4.64c}
$$

Note, that the accuracy in the latter expressions (4.64a–c) differs from that in the expressions for the means (4.63a–c). This is because the formulae (4.64a–c) constitute corrections to the equations for long-wavelength fluctuations (4.50a and b), resulting from interactions of the short-wavelength ones. In the following steps of the renormalization procedure they will contribute to the means (4.63a–c), which constitute the corrections to the mean-field equations (4.32a and b). It is, therefore, enough to calculate only the first non-vanishing order of the fluctuational corrections (4.64a–c), which will then lead to the final expressions for the mean EMF and the effective Lorentz force in turbulence, when the nonlinear evolution of turbulent fluctuations is considered.

We stress, that the second part of the distant interactions limit (4.57) involving $\omega/\omega' \to 0$ has not been taken yet, since we still need to properly take into account the frequency shifts $\delta(\omega \pm \tilde{\omega})$ from non-stationary force correlations, in the limit

$$
\tilde{\omega}t \ll 1, \quad \tilde{\omega} \ll \nu K_e^2,\ \eta K_e^2. \tag{4.65}
$$

These have to be treated in the distributional sense, i.e. we multiply the equations for the Fourier transforms by $\exp(-\mathrm{i}\omega t)$ and integrate over all frequencies, which in the case of mean field equations yields for the corrections (4.63a–c)

$$
\begin{aligned}
\int_{-\infty}^{\infty} f(\omega)(\delta(\omega+\tilde{\omega})-\delta(\omega-\tilde{\omega}))\mathrm{e}^{-\mathrm{i}\omega t}\mathrm{d}\omega &= f(-\tilde{\omega})\mathrm{e}^{\mathrm{i}\tilde{\omega}t} - f(\tilde{\omega})\mathrm{e}^{-\mathrm{i}\tilde{\omega}t}\\
&\approx 2\mathrm{i}\tilde{\omega}tf(0) - 2\tilde{\omega}\frac{\mathrm{d}f}{\mathrm{d}\omega}\bigg|_{\omega=0} + \mathcal{O}(\tilde{\omega}^2t^2),
\end{aligned}
\tag{4.66}
$$

for any function $f(\omega)$. In the case of the fluctuational correction (4.64c) we get

$$
\begin{aligned}
&\int_{-\infty}^{\infty}\frac{\hat{b}^{<}_p(\omega+\tilde{\omega},\mathbf{k})}{\gamma_\eta(\omega-\omega',k')\gamma_\nu(-\omega'-\tilde{\omega},k')}\mathrm{e}^{-\mathrm{i}\omega t}\mathrm{d}\omega\\
&=\int_{-\infty}^{\infty}\frac{\hat{b}^{<}_p(\omega,\mathbf{k})\mathrm{e}^{\mathrm{i}\tilde{\omega}t}}{\gamma_\eta(\omega-\tilde{\omega}-\omega',k')\gamma_\nu(-\omega'-\tilde{\omega},k')}\mathrm{e}^{-\mathrm{i}\omega t}\mathrm{d}\omega,
\end{aligned}
\tag{4.67a}
$$

$$\begin{aligned}&\int_{-\infty}^{\infty}\frac{\hat{b}_p^{<}(\omega-\tilde{\omega},\mathbf{k})}{\gamma_\eta(\omega-\omega',k')\gamma_\nu(-\omega'+\tilde{\omega},k')}\mathrm{e}^{-\mathrm{i}\omega t}\mathrm{d}\omega\\&=\int_{-\infty}^{\infty}\frac{\hat{b}_p^{<}(\omega,\mathbf{k})\mathrm{e}^{-\mathrm{i}\tilde{\omega}t}}{\gamma_\eta(\omega+\tilde{\omega}-\omega',k')\gamma_\nu(-\omega'+\tilde{\omega},k')}\mathrm{e}^{-\mathrm{i}\omega t}\mathrm{d}\omega,\end{aligned}\tag{4.67b}$$

and then we return back to the Fourier space (we take off the $\mathrm{d}\omega$ integral and the factor $\mathrm{e}^{-\mathrm{i}\omega t}$), and take the distant-interaction limit $\omega/\omega'\to 0$. Since

$$\begin{aligned}&\frac{\mathrm{e}^{\mathrm{i}\tilde{\omega}t}}{\gamma_\eta(-\tilde{\omega}-\omega',k')\gamma_\nu(-\omega'-\tilde{\omega},k')}-\frac{\mathrm{e}^{-\mathrm{i}\tilde{\omega}t}}{\gamma_\eta(\tilde{\omega}-\omega',k')\gamma_\nu(-\omega'+\tilde{\omega},k')}\\&=\frac{2\mathrm{i}\tilde{\omega}t}{\gamma_\eta(-\omega',k')\gamma_\nu(-\omega',k')}-2\tilde{\omega}\frac{\mathrm{d}}{\mathrm{d}w}\frac{1}{\gamma_\eta(w-\omega',k')\gamma_\nu(-\omega'+w,k')}\Bigg|_{w=0}\\&+\mathcal{O}(\tilde{\omega}^2t^2),\end{aligned}\tag{4.68}$$

this procedure finally yields the following substitution in (4.64c)

$$\begin{aligned}&\frac{1}{\gamma_\eta(\omega-\omega',k')}\left[\frac{\hat{b}_p^{<}(\omega+\tilde{\omega},\mathbf{k})}{\gamma_\nu(-\omega'-\tilde{\omega},k')}-\frac{\hat{b}_p^{<}(\omega-\tilde{\omega},\mathbf{k})}{\gamma_\nu(-\omega'+\tilde{\omega},k')}\right]\\&\longmapsto\frac{2\mathrm{i}\tilde{\omega}t\hat{b}_p^{<}(\omega,\mathbf{k})}{\gamma_\eta(-\omega',k')\gamma_\nu(-\omega',k')}-2\tilde{\omega}\hat{b}_p^{<}(\omega,\mathbf{k})\frac{\mathrm{d}}{\mathrm{d}w}\frac{1}{\gamma_\eta(w-\omega',k')\gamma_\nu(-\omega'+w,k')}\Bigg|_{w=0}.\end{aligned}\tag{4.69}$$

It follows, that the mean corrections (4.63a–c) are transformed into

$$\begin{aligned}&-\varepsilon Ro\frac{\partial}{\partial X_j}\left\langle\theta_{K_\nu}\int\mathrm{d}^4q\mathrm{e}^{\mathrm{i}(\mathbf{k}\cdot\mathbf{x}-\omega t)}\int\mathrm{d}^4q'\hat{u}_i^{>}(\mathbf{q}')\hat{u}_j^{>}(\mathbf{q}-\mathbf{q}')\right\rangle\\&=\varepsilon\left(4\frac{\partial}{\partial X_i}\left(\frac{\langle B\rangle^2}{2}\right)-(\langle B\rangle\cdot\bar{\nabla})\langle B\rangle_i\right)\frac{8\pi}{15}Ro\left(Ro\tilde{\beta}\right)^2HD_0\theta_{K_\nu}\times\\&\int_{k_1}^{K_\nu}k'\mathrm{d}k'\{(1+\Gamma\tilde{\omega}t)[\mathcal{I}_5(k')+\mathcal{I}_7(k')]-\Gamma\tilde{\omega}[2\mathcal{I}_8(k')+\mathcal{I}_9(k')+\mathcal{I}_{10}(k')]\}\\&+\mathcal{O}\left(Ro\left(Ro\tilde{\beta}\right)^4\right),\end{aligned}\tag{4.70a}$$

$$\begin{aligned}&\varepsilon RoH\frac{\partial}{\partial X_j}\left\langle\theta_{K_\eta}\int\mathrm{d}^4q\mathrm{e}^{\mathrm{i}(\mathbf{k}\cdot\mathbf{x}-\omega t)}\int\mathrm{d}^4q'\hat{b}_i^{>}(\mathbf{q}')\hat{b}_j^{>}(\mathbf{q}-\mathbf{q}')\right\rangle\\&=\varepsilon\left(4\frac{\partial}{\partial X_i}\left(\frac{\langle B\rangle^2}{2}\right)-(\langle B\rangle\cdot\bar{\nabla})\langle B\rangle_i\right)\frac{8\pi}{15}Ro\left(Ro\tilde{\beta}\right)^2HD_0\theta_{K_\eta}\times\\&\int_{k_1}^{K_\eta}k'\mathrm{d}k'\{(1+\Gamma\tilde{\omega}t)\mathcal{I}_1(k')-\Gamma\tilde{\omega}[\mathcal{I}_2(k')+\mathcal{I}_3(k')]\}+\mathcal{O}\left(Ro\left(Ro\tilde{\beta}\right)^4\right),\end{aligned}\tag{4.70b}$$

$$\frac{1}{2}\varepsilon RoH\frac{\partial}{\partial X_i}\left\langle\theta_{K_\eta}\int \mathrm{d}^4q\mathrm{e}^{\mathrm{i}(\mathbf{k}\cdot\mathbf{x}-\omega t)}\int \mathrm{d}^4q'\hat{b}_j^>(\mathbf{q}')\hat{b}_j^>(\mathbf{q}-\mathbf{q}')\right\rangle$$
$$=\varepsilon\frac{\partial}{\partial X_i}\left(\frac{\langle B\rangle^2}{2}\right)\frac{8\pi}{3}Ro\left(Ro\tilde{\beta}\right)^2HD_0\theta_{K_\eta}\times \tag{4.70c}$$
$$\int_{k_1}^{K_\eta}k'\mathrm{d}k'\{(1+\Gamma\tilde{\omega}t)\mathcal{I}_1(k')-\Gamma\tilde{\omega}[\mathcal{I}_2(k')+\mathcal{I}_3(k')]\}+\mathcal{O}\left(Ro\left(Ro\tilde{\beta}\right)^4\right),$$

$$\varepsilon Ro\tilde{\beta}^{-1}\epsilon_{ijk}\frac{\partial}{\partial X_j}\epsilon_{kmn}\left\langle\theta_K\int \mathrm{d}^4q\mathrm{e}^{\mathrm{i}(\mathbf{k}\cdot\mathbf{x}-\omega t)}\int \mathrm{d}^4q'\hat{u}_m^>(\mathbf{q}')\hat{b}_n^>(\mathbf{q}-\mathbf{q}')\right\rangle$$
$$=\varepsilon(\bar{\nabla}\times\langle\mathbf{B}\rangle)_i\frac{8\pi}{3}Ro^2D_1\theta_K\int_{k_1}^{K}\frac{\mathrm{d}k'}{k'}\{(1+\Gamma\tilde{\omega}t)\mathcal{I}_4(k')-\Gamma\tilde{\omega}[\mathcal{I}_5(k')+\mathcal{I}_6(k')]\}$$
$$-\varepsilon[\bar{\nabla}\times(\langle B\rangle^2\langle\mathbf{B}\rangle)]_i\frac{8\pi}{5}Ro^2\left(Ro\tilde{\beta}\right)^2HD_1\theta_K\int_{k_1}^{K}k'\mathrm{d}k'\{(1+\Gamma\tilde{\omega}t)[\mathcal{I}_9(k') \tag{4.70d}$$
$$+\mathcal{I}_{11}(k')]-\Gamma\tilde{\omega}[2\mathcal{I}_{12}(k')+2\mathcal{I}_{13}(k')+\mathcal{I}_{14}(k')+\mathcal{I}_{15}(k')]\}$$
$$+\mathcal{O}\left(Ro^2\left(Ro\tilde{\beta}\right)^4\right),$$

and for the fluctuations the only non-negligible correction takes the form

$$\mathrm{i}Ro\theta_K\epsilon_{ijk}\epsilon_{kmn}k_j\int \mathrm{d}^4q'\left[\left\langle\hat{u}_m^>(\mathbf{q}')\hat{b}_n^>(\mathbf{q}-\mathbf{q}')\right\rangle_c-\left\langle\hat{u}_m^>(\mathbf{q}')\hat{b}_n^>(\mathbf{q}-\mathbf{q}')\right\rangle\right]$$
$$=\epsilon_{ijk}\mathrm{i}k_j\hat{b}_k^<(\omega,\mathbf{k})\frac{8\pi}{3}Ro^2D_1\theta_K\int_{k_1}^{K}\frac{\mathrm{d}k'}{k'}\{(1+\Gamma\tilde{\omega}t)\mathcal{I}_4(k')-\Gamma\tilde{\omega}[\mathcal{I}_5(k')+\mathcal{I}_6(k')]\} \tag{4.71}$$
$$+\mathcal{O}\left(Ro^2\left(Ro\tilde{\beta}\right)\right),$$

where

$$\mathcal{I}_1(k')=\int_{-\infty}^{\infty}\frac{d\omega'}{\left(\omega'^2+E_\nu^2k'^4\right)\left(\omega'^2+E_\eta^2k'^4\right)}=\frac{\pi}{E_\nu E_\eta\left(E_\nu+E_\eta\right)k'^6}, \tag{4.72a}$$

$$\mathcal{I}_2(k')=\int_{-\infty}^{\infty}\frac{(-\mathrm{i}\omega'+E_\nu k'^2)\mathrm{d}\omega'}{\left(\omega'^2+E_\nu^2k'^4\right)^2\left(\omega'^2+E_\eta^2k'^4\right)}=\frac{\pi\left(2E_\nu+E_\eta\right)}{2E_\nu^2E_\eta\left(E_\nu+E_\eta\right)^2k'^8}, \tag{4.72b}$$

$$\mathcal{I}_3(k')=\int_{-\infty}^{\infty}\frac{\left(-\mathrm{i}\omega'+E_\eta k'^2\right)\mathrm{d}\omega'}{\left(\omega'^2+E_\nu^2k'^4\right)\left(\omega'^2+E_\eta^2k'^4\right)^2}=\frac{\pi\left(E_\nu+2E_\eta\right)}{2E_\nu E_\eta^2\left(E_\nu+E_\eta\right)^2k'^8}, \tag{4.72c}$$

$$\mathcal{I}_4(k')=\int_{-\infty}^{\infty}\frac{\mathrm{d}\omega'}{\left(\omega'^2+E_\nu^2k'^4\right)\left(\mathrm{i}\omega'+E_\eta k'^2\right)}=\frac{\pi}{E_\nu\left(E_\nu+E_\eta\right)k'^4}, \tag{4.72d}$$

$$\mathcal{I}_5(k') = \mathcal{I}_7(k') = \int_{-\infty}^{\infty} \frac{\left(-\mathrm{i}\omega' + E_\nu k'^2\right)\mathrm{d}\omega'}{\left(\omega'^2 + E_\nu^2 k'^4\right)^2\left(\mathrm{i}\omega' + E_\eta k'^2\right)} = \frac{\pi}{2E_\nu^2\left(E_\nu + E_\eta\right)k'^6}, \quad (4.72e)$$

$$\mathcal{I}_6(k') = \int_{-\infty}^{\infty} \frac{\mathrm{d}\omega'}{\left(\omega'^2 + E_\nu^2 k'^4\right)\left(\mathrm{i}\omega' + E_\eta k'^2\right)^2} = \frac{\pi}{E_\nu\left(E_\nu + E_\eta\right)^2 k'^6}, \quad (4.72f)$$

$$\mathcal{I}_8(k') = \int_{-\infty}^{\infty} \frac{\left(-\mathrm{i}\omega' + E_\nu k'^2\right)^2\mathrm{d}\omega'}{\left(\omega'^2 + E_\nu^2 k'^4\right)^3\left(\mathrm{i}\omega' + E_\eta k'^2\right)} = \frac{\pi}{4E_\nu^3\left(E_\nu + E_\eta\right)k'^8}, \quad (4.72g)$$

$$\mathcal{I}_9(k') = \int_{-\infty}^{\infty} \frac{\left(-\mathrm{i}\omega' + E_\nu k'^2\right)\mathrm{d}\omega'}{\left(\omega'^2 + E_\nu^2 k'^4\right)^2\left(\mathrm{i}\omega' + E_\eta k'^2\right)^2} = \frac{\pi}{2E_\nu^2\left(E_\nu + E_\eta\right)^2 k'^8}, \quad (4.72h)$$

$$\mathcal{I}_{10}(k') = \int_{-\infty}^{\infty} \frac{\mathrm{d}\omega'}{\left(\omega'^2 + E_\nu^2 k'^4\right)^2\left(-\mathrm{i}\omega' + E_\eta k'^2\right)} = \frac{\pi\left(2E_\nu + E_\eta\right)}{2E_\nu^3\left(E_\nu + E_\eta\right)^2 k'^8}, \quad (4.72i)$$

$$\mathcal{I}_{11}(k') = \int_{-\infty}^{\infty} \frac{\left(\mathrm{i}\omega' + E_\nu k'^2\right)\mathrm{d}\omega'}{\left(\omega'^2 + E_\nu^2 k'^4\right)^2\left(\omega'^2 + E_\eta^2 k'^4\right)} = \frac{\pi\left(2E_\nu + E_\eta\right)}{2E_\nu^2 E_\eta\left(E_\nu + E_\eta\right)^2 k'^8}, \quad (4.72j)$$

$$\mathcal{I}_{12}(k') = \int_{-\infty}^{\infty} \frac{\left(-\mathrm{i}\omega' + E_\nu k'^2\right)^2\mathrm{d}\omega'}{\left(\omega'^2 + E_\nu^2 k'^4\right)^3\left(\mathrm{i}\omega' + E_\eta k'^2\right)^2} = \frac{\pi}{4E_\nu^3\left(E_\nu + E_\eta\right)^2 k'^{10}}, \quad (4.72k)$$

$$\mathcal{I}_{13}(k') = \int_{-\infty}^{\infty} \frac{\left(-\mathrm{i}\omega' + E_\nu k'^2\right)\mathrm{d}\omega'}{\left(\omega'^2 + E_\nu^2 k'^4\right)^2\left(\mathrm{i}\omega' + E_\eta k'^2\right)^3} = \frac{\pi}{2E_\nu^2\left(E_\nu + E_\eta\right)^3 k'^{10}}, \quad (4.72l)$$

$$\mathcal{I}_{14}(k') = \int_{-\infty}^{\infty} \frac{\mathrm{d}\omega'}{\left(\omega'^2 + E_\nu^2 k'^4\right)^2\left(\omega'^2 + E_\eta^2 k'^4\right)} = \frac{\pi\left(2E_\nu + E_\eta\right)}{2E_\nu^3 E_\eta\left(E_\nu + E_\eta\right)^2 k'^{10}}, \quad (4.72m)$$

$$\mathcal{I}_{15}(k') = \int_{-\infty}^{\infty} \frac{\left(\mathrm{i}\omega' + E_\nu k'^2\right)\left(-\mathrm{i}\omega' + E_\eta k'^2\right)\mathrm{d}\omega'}{\left(\omega'^2 + E_\nu^2 k'^4\right)^2\left(\omega'^2 + E_\eta^2 k'^4\right)^2} = \frac{\pi\left(E_\nu^2 + 4E_\nu E_\eta + E_\eta^2\right)}{2E_\nu^2 E_\eta^2\left(E_\nu + E_\eta\right)^3 k'^{10}}. \quad (4.72n)$$

Ultimately, we are interested in the long-wavelength limit for fluctuations $k \to K_e^+$, because in this limit the contributions to the dynamics of the mean fields from the fluctuating fields is greatest. Since we assume $K_e \ll K_\nu$, K_η the dependence of the mean EMF and the mean Lorentz force on the short-wavelength cut-offs K_ν and K_η

is lost in this limit (similarly as in [8, 18]). Consequently, in the following, we may slightly facilitate the notation by taking $\theta_{K_\nu} = 1$, $\theta_{K_\eta} = 1$ and $\theta_K = 1$, as these factors become unity as soon as the wave number falls below $K = \min(K_\nu, K_\eta)$.

We now utilize the assumption of narrowness of the first spectral bite $K_{\max} - k_1 = \delta k \ll 1$ and define the following coefficients which describe the average effect of the short-wavelength fluctuations with wave numbers from the narrow band $k_1 \leqslant k \leqslant K_{\max}$ on the means

$$\begin{aligned}\bar{Q}(k_1) = 1 - \frac{8\pi^2}{15}\frac{Ro^2}{E_\nu^2 E_\eta}\frac{D_0\delta k}{k_1^5}\Bigg[&\frac{\left(\theta_{K_\eta}E_\nu + \theta_{K_\nu}E_\eta\right)}{E_\nu + E_\eta}(1 + \Gamma\tilde{\varpi}t)\\ &- \frac{\theta_{K_\nu}\left(4E_\nu E_\eta^2 + 2E_\eta^3\right) + \theta_{K_\eta}\left(E_\nu^3 + 4E_\nu^2 E_\eta + E_\nu E_\eta^2\right)}{E_\nu E_\eta\left(E_\nu + E_\eta\right)^2}\frac{\Gamma\tilde{\varpi}}{2k_1^2}\Bigg],\end{aligned} \tag{4.73a}$$

$$\begin{aligned}\bar{Q}_p(k_1) = 1 + \frac{8\pi^2}{15}\frac{Ro^2}{E_\nu^2 E_\eta\left(E_\nu + E_\eta\right)}\frac{D_0\delta k}{k_1^5}\Bigg[&\left(\theta_{K_\eta}E_\nu - 4\theta_{K_\nu}E_\eta\right)(1 + \Gamma\tilde{\varpi}t)\\ &- \frac{\theta_{K_\eta}\left(E_\nu^3 + 4E_\nu^2 E_\eta + E_\nu E_\eta^2\right) - \theta_{K_\nu}\left(16E_\nu E_\eta^2 + 8E_\eta^3\right)}{E_\nu E_\eta\left(E_\nu + E_\eta\right)}\frac{\Gamma\tilde{\varpi}}{2k_1^2}\Bigg],\end{aligned} \tag{4.73b}$$

$$\bar{\alpha}_A(k_1) = \frac{8\pi^2}{3}\frac{Ro^2\theta_K}{E_\nu\left(E_\nu + E_\eta\right)}\frac{D_1\delta k}{k_1^5}\left[1 + \Gamma\tilde{\varpi}t - \frac{3E_\nu + E_\eta}{E_\nu\left(E_\nu + E_\eta\right)}\frac{\Gamma\tilde{\varpi}}{2k_1^2}\right], \tag{4.73c}$$

$$\bar{\alpha}_S(k_1) = \frac{8\pi^2}{5}\frac{Ro^2\left(Ro\tilde{\beta}\right)^2 H\theta_K}{E_\nu^2 E_\eta\left(E_\nu + E_\eta\right)}\frac{D_1\delta k}{k_1^7}\left[1 + \Gamma\tilde{\varpi}t - \frac{E_\nu^2 + 5E_\nu E_\eta + 2E_\eta^2}{E_\nu E_\eta\left(E_\nu + E_\eta\right)}\frac{\Gamma\tilde{\varpi}}{2k_1^2}\right], \tag{4.73d}$$

and a coefficient describing the effect of short-wavelength fluctuations on the long-wavelength fluctuations corresponding to the band $K_e \leqslant k < k_1$

$$\breve{\alpha}(k_1) = \frac{8\pi^2}{3}\frac{Ro^2\theta_K}{E_\nu\left(E_\nu + E_\eta\right)}\frac{D_1\delta k}{k_1^5}\left[(1 + \Gamma\tilde{\varpi}t) - \frac{3E_\nu + E_\eta}{E_\nu\left(E_\nu + E_\eta\right)}\frac{\Gamma\tilde{\varpi}}{2k_1^2}\right]. \tag{4.74}$$

With the use of those definitions we can write down the dynamical equations (4.32a and b) and (4.50a and b) in the new form, with the effect of the short-wavelength modes $\mathbf{u}^>$, $\mathbf{b}^>$ expressed through the effective EMF and the Lorentz force

$$\begin{aligned}\frac{\partial\langle\mathbf{U}\rangle}{\partial t} + \varepsilon Ro\left(\langle\mathbf{U}\rangle\cdot\bar{\nabla}\right)\langle\mathbf{U}\rangle = &- \varepsilon\bar{\nabla}\left(\frac{\langle p\rangle}{\rho} + Ro\tilde{\beta}^2 H\bar{Q}_p(k_1)\frac{\langle B\rangle^2}{2}\right)\\ &+ \varepsilon Ro\tilde{\beta}^2 H\bar{Q}(k_1)\left(\langle\mathbf{B}\rangle\cdot\bar{\nabla}\right)\langle\mathbf{B}\rangle\\ &+ E_\nu\varepsilon^2\bar{\nabla}^2\langle\mathbf{U}\rangle - \varepsilon Ro\left(\bar{\nabla}\cdot\langle\mathbf{u}^<\mathbf{u}^<\rangle - H\bar{\nabla}\cdot\langle\mathbf{b}^<\mathbf{b}^<\rangle\right),\end{aligned} \tag{4.75a}$$

$$\begin{aligned}\frac{\partial\langle\mathbf{B}\rangle}{\partial t} &= \varepsilon\bar{\nabla}\times\left[\left(\bar{\alpha}_A(k_1)-\bar{\alpha}_S(k_1)\langle B\rangle^2\right)\langle\mathbf{B}\rangle\right]+\varepsilon Ro\tilde{\beta}^{-1}\bar{\nabla}\times\langle\mathbf{u}^<\times\mathbf{b}^<\rangle\\ &+\varepsilon Ro\bar{\nabla}\times(\langle\mathbf{U}\rangle\times\langle\mathbf{B}\rangle)+\varepsilon^2E_\eta\bar{\nabla}^2\langle\mathbf{B}\rangle,\end{aligned}\tag{4.75b}$$

$$\begin{aligned}(-\mathrm{i}\omega+E_\nu k^2)\hat{u}_i^<(\mathbf{q}) &= \hat{f}_i^<(\mathbf{q})+\mathrm{i}Ro\tilde{\beta}H(\mathbf{k}\cdot\langle\mathbf{B}\rangle)\hat{b}_i^<(\mathbf{q})\\ &-\frac{1}{2}\mathrm{i}RoP_{imn}(\mathbf{k})\left[\mathbb{I}_{mn}^{(u^<)}-H\mathbb{I}_{mn}^{(b^<)}\right],\end{aligned}\tag{4.75c}$$

$$\left(-\mathrm{i}\omega+E_\eta k^2\right)\hat{b}_i^<(\mathbf{q})-\breve{\alpha}(k_1)\epsilon_{ijk}\mathrm{i}k_j\hat{b}_k^< = \mathrm{i}Ro\tilde{\beta}\left(\mathbf{k}\cdot\langle\mathbf{B}\rangle\right)\hat{u}_i^<(\mathbf{q})+\mathrm{i}Rok_j\mathbb{I}_{ij}^{(u^<b^<)}.\tag{4.75d}$$

We now proceed to the next step of the renormalization procedure which consists of a step-by-step elimination of infinitesimally small wave-number bands from the Fourier spectrum from the short-wavelength side. We introduce k_2, which satisfies

$$\delta k = k_1 - k_2 \ll 1,\tag{4.76}$$

and, again, split the remaining fluctuational Fourier spectrum $K_e \leqslant k \leqslant k_1$ into two parts by defining new variables (but keeping the same notation)

$$\theta(k-k_2)\hat{u}_i^<(\mathbf{k},\omega)\mapsto\hat{u}_i^>(\mathbf{k},\omega),\qquad(\text{for } k_2<k<k_1),\tag{4.77}$$

$$\theta(k_2-k)\hat{u}_i^<(\mathbf{k},\omega)\mapsto\hat{u}_i^<(\mathbf{k},\omega),\qquad(\text{for } k<k_2),\tag{4.78}$$

and the same way for $\hat{\mathbf{b}}$ and $\hat{\mathbf{f}}$. The equations for fluctuations are also split, as in the first step (cf (4.50a and b) and (4.51a and b)), i.e. we have

$$\begin{aligned}&(-\mathrm{i}\omega+E_\nu k^2)\hat{u}_i^<(\mathbf{q})=\hat{f}_i^<(\mathbf{q})+\mathrm{i}Ro\tilde{\beta}H(\mathbf{k}\cdot\langle\mathbf{B}\rangle)\hat{b}_i^<(\mathbf{q})\\ &-\frac{1}{2}\mathrm{i}RoP_{imn}(\mathbf{k})[\mathbb{I}_{mn}^{(u^<)}-H\mathbb{I}_{mn}^{(b^<)}-\langle\mathbb{I}_{mn}^{(u^<)}\rangle+H\langle\mathbb{I}_{mn}^{(b^<)}\rangle]\\ &-\frac{1}{2}\mathrm{i}RoP_{imn}(\mathbf{k})\Bigg\{\theta_{K_\nu}\int_{K_e}^{K_\nu}d^4q'\left[\langle\hat{u}_m^>(\mathbf{q}')\hat{u}_n^>(\mathbf{q}-\mathbf{q}')\rangle_c-\langle\hat{u}_m^>(\mathbf{q}')\hat{u}_n^>(\mathbf{q}-\mathbf{q}')\rangle\right]\\ &\qquad-H\theta_{K_\eta}\int_{K_e}^{K_\eta}d^4q'\left[\left\langle\hat{b}_m^>(\mathbf{q}')\hat{b}_n^>(\mathbf{q}-\mathbf{q}')\right\rangle_c-\left\langle\hat{b}_m^>(\mathbf{q}')\hat{b}_n^>(\mathbf{q}-\mathbf{q}')\right\rangle\right]\Bigg\},\end{aligned}\tag{4.79a}$$

$$\begin{aligned}&\left(-\mathrm{i}\omega+E_\eta k^2\right)\hat{b}_i^<(\mathbf{q})-\breve{\alpha}(k_1)\epsilon_{ijk}\mathrm{i}k_j\hat{b}_k^<\\ &=\mathrm{i}Ro\tilde{\beta}(\mathbf{k}\cdot\langle\mathbf{B}\rangle)\hat{u}_i^<(\mathbf{q})+\mathrm{i}Rok_j\left[\mathbb{I}_{ij}^{(u^<b^<)}-\left\langle\mathbb{I}_{ij}^{(u^<b^<)}\right\rangle\right]\\ &+\mathrm{i}Rok_j\epsilon_{ijk}\epsilon_{kmn}\theta_K\int_{K_e}^{K}d^4q'\left[\left\langle\hat{u}_m^>(\mathbf{q}')\hat{b}_n^>(\mathbf{q}-\mathbf{q}')\right\rangle_c-\left\langle\hat{u}_m^>(\mathbf{q}')\hat{b}_n^>(\mathbf{q}-\mathbf{q}')\right\rangle\right],\end{aligned}\tag{4.79b}$$

for the new long-wavelength modes and

$$
\begin{aligned}
\hat{u}_i^{>}(\mathbf{q}) = {} & \frac{1}{\gamma_u}\hat{f}_i^{>}(\mathbf{q}) - \frac{1}{2}\mathrm{i}Ro\frac{P_{imn}(\mathbf{k})}{\gamma_u}[\mathbb{I}^{(u^<)}_{mn} - H\mathbb{I}^{(b^<)}_{mn} - \langle\mathbb{I}^{(u^<)}_{mn}\rangle + H\langle\mathbb{I}^{(b^<)}_{mn}\rangle] \\
& - HRo\left(Ro\tilde{\beta}\right)\frac{\mathbf{k}\cdot\langle\mathbf{B}\rangle}{\gamma_u\gamma_\eta}k_j\left[\mathbb{I}^{(u^<b^<)}_{ij} - \left\langle\mathbb{I}^{(u^<b^<)}_{ij}\right\rangle\right] \\
& - \mathrm{i}\frac{Ro}{\gamma_u}P_{imn}[\mathbb{J}^{(u)}_{mn} - H\mathbb{J}^{(b)}_{mn}] - HRo\left(Ro\tilde{\beta}\right)\frac{\mathbf{k}\cdot\langle\mathbf{B}\rangle}{\gamma_u\gamma_\eta}k_j\mathbb{J}^{(ub)}_{ij} + R_i^{(u)},
\end{aligned}
\tag{4.80a}
$$

$$
\begin{aligned}
\hat{b}_i^{>}(\mathbf{q}) = {} & \mathrm{i}Ro\tilde{\beta}\frac{\mathbf{k}\cdot\langle\mathbf{B}\rangle}{\gamma_u\gamma_\eta}\left[\delta_{ij} + \mathrm{i}\frac{\breve{\alpha}(k_1)}{\gamma_\eta}\epsilon_{ilj}k_l\right]\hat{f}_j^{>}(\mathbf{q}) + \mathrm{i}\frac{Ro}{\gamma_\eta}k_j\left[\mathbb{I}^{(u^<b^<)}_{ij} - \left\langle\mathbb{I}^{(u^<b^<)}_{ij}\right\rangle\right] \\
& + \frac{1}{2}Ro\left(Ro\tilde{\beta}\right)\frac{\mathbf{k}\cdot\langle\mathbf{B}\rangle}{\gamma_u\gamma_\eta}P_{imn}(\mathbf{k})[\mathbb{I}^{(u^<)}_{mn} - H\mathbb{I}^{(b^<)}_{mn} - \langle\mathbb{I}^{(u^<)}_{mn}\rangle + H\langle\mathbb{I}^{(b^<)}_{mn}\rangle] \\
& + \mathrm{i}\frac{Ro}{\gamma_\eta}k_j\mathbb{J}^{(ub)}_{ij} + Ro\left(Ro\tilde{\beta}\right)\frac{\mathbf{k}\cdot\langle\mathbf{B}\rangle}{\gamma_u\gamma_\eta}P_{imn}[\mathbb{J}^{(u)}_{mn} - H\mathbb{J}^{(b)}_{mn}] + R_i^{(b)},
\end{aligned}
\tag{4.80b}
$$

for the new short-wavelength ones. Of course now $\langle\cdot\rangle_c$ denotes conditional average over the second shell ($k_2 \leqslant k \leqslant k_1$) statistical subensemble.

Repetition of the sub-steps undertaken in the first step of the renormalization procedure, but with the new term proportional to $\breve{\alpha}(k_1) \sim Ro^2$ in the equation for the short-wavelength magnetic fluctuation (4.80b), leads to modification of the mean EMF and the mean Lorentz force. However, the coefficient $\breve{\alpha}(k_2)$ remains unaltered at leading order with respect to $\breve{\alpha}(k_1)$ obtained from the first step. The $\breve{\alpha}(k_1)$-term in the equation (4.80b) generates corrections of the order $\mathcal{O}(Ro^3(Ro\tilde{\beta})^2)$ in the mean Lorentz force (4.70a–c) and of the order $\mathcal{O}(Ro^4)$ and $\mathcal{O}(Ro^4(Ro\tilde{\beta})^2)$ in the mean EMF (4.70d). We neglect the former, that is terms of the order $\mathcal{O}(Ro^3(Ro\tilde{\beta})^2)$ and consider only the largest corrections in the mean EMF, that is those of the order $\mathcal{O}(Ro^4)$. Consequently, only the mean coefficient $\bar{\alpha}_A$ is modified in the second step of the procedure, i.e.

$$
\begin{aligned}
\bar{\alpha}_A(k_2) = {} & \bar{\alpha}_A(k_1) + \frac{8\pi^2}{3}\frac{Ro^2\theta_K\delta k}{E_\nu\left(E_\nu + E_\eta\right)}\left\{(1 + \Gamma\tilde{\omega}t)\left[\frac{D_1}{k_2^5} - \breve{\alpha}(k_2)\frac{D_0}{\left(E_\nu + E_\eta\right)k_2^5}\right]\right. \\
& \left. - \frac{\Gamma\tilde{\omega}}{2E_\nu\left(E_\nu + E_\eta\right)}\left[\left(3E_\nu + E_\eta\right)\frac{D_1}{k_2^7} - \breve{\alpha}(k_2)\frac{\left(5E_\nu + E_\eta\right)D_0}{\left(E_\nu + E_\eta\right)k_2^7}\right]\right\}.
\end{aligned}
\tag{4.81}
$$

It is clear now, that following steps of the renormalization procedure do not introduce any new terms of the order $\mathcal{O}(Ro^2(Ro\tilde{\beta})^2, Ro^4)$ or greater, in the expressions for the coefficients $\bar{Q}$, $\bar{Q}_p$, $\bar{\alpha}_A$, $\bar{\alpha}_S$, nor any terms of the order $\mathcal{O}(Ro^2)$ or greater in $\breve{\alpha}$. In the limit $\delta k \to 0$, $\tilde{\omega}t \ll 1$ the equations (4.73a and b), (4.81), (4.73d) and (4.74) transform into the following set of recursion differential equations

$$\frac{\mathrm{d}\bar{Q}}{\mathrm{d}k} = \frac{8\pi^2}{15}\frac{Ro^2}{E_\nu^2 E_\eta}\frac{D_0}{k^5}\left[\frac{\left(\theta_{K_\eta}E_\nu + \theta_{K_\nu}E_\eta\right)}{E_\nu + E_\eta} - \frac{\theta_{K_\nu}\left(4E_\nu E_\eta^2 + 2E_\eta^3\right) + \theta_{K_\eta}\left(E_\nu^3 + 4E_\nu^2 E_\eta + E_\nu E_\eta^2\right)}{E_\nu E_\eta\left(E_\nu + E_\eta\right)^2}\frac{\Gamma\tilde{\omega}}{2k^2}\right], \tag{4.82a}$$

$$\frac{\mathrm{d}\bar{Q}_p}{\mathrm{d}k} = -\frac{8\pi^2}{15}\frac{Ro^2}{E_\nu^2 E_\eta\left(E_\nu + E_\eta\right)}\frac{D_0}{k^5}\left[\left(\theta_{K_\eta}E_\nu - 4\theta_{K_\nu}E_\eta\right) - \frac{\theta_{K_\eta}\left(E_\nu^3 + 4E_\nu^2 E_\eta + E_\nu E_\eta^2\right) - \theta_{K_\nu}\left(16E_\nu E_\eta^2 + 8E_\eta^3\right)}{E_\nu E_\eta\left(E_\nu + E_\eta\right)}\frac{\Gamma\tilde{\omega}}{2k^2}\right], \tag{4.82b}$$

$$\frac{\mathrm{d}\bar{\alpha}_A}{\mathrm{d}k} = -\frac{8\pi^2}{3}\frac{Ro^2\theta_K}{E_\nu\left(E_\nu + E_\eta\right)}\left\{\frac{D_1}{k^5} - \frac{D_0\breve{\alpha}(k)}{\left(E_\nu + E_\eta\right)k^5} - \frac{\Gamma\tilde{\omega}}{2E_\nu\left(E_\nu + E_\eta\right)}\left[\left(3E_\nu + E_\eta\right)\frac{D_1}{k^7} - \breve{\alpha}(k)\frac{\left(5E_\nu + E_\eta\right)D_0}{\left(E_\nu + E_\eta\right)k^7}\right]\right\}, \tag{4.82c}$$

$$\frac{\mathrm{d}\bar{\alpha}_S}{\mathrm{d}k} = -\frac{8\pi^2}{5}\frac{Ro^2\left(Ro\tilde{\beta}\right)^2 H\theta_K}{E_\nu^2 E_\eta\left(E_\nu + E_\eta\right)}\frac{D_1}{k^7}\left(1 - \frac{E_\nu^2 + 5E_\nu E_\eta + 2E_\eta^2}{E_\nu E_\eta\left(E_\nu + E_\eta\right)}\frac{\Gamma\tilde{\omega}}{2k^2}\right), \tag{4.82d}$$

$$\frac{\mathrm{d}\breve{\alpha}}{\mathrm{d}k} = -\frac{8\pi^2}{3}\frac{Ro^2\theta_K}{E_\nu\left(E_\nu + E_\eta\right)}\frac{D_1}{k^5}\left(1 - \frac{3E_\nu + E_\eta}{E_\nu\left(E_\nu + E_\eta\right)}\frac{\Gamma\tilde{\omega}}{2k^2}\right). \tag{4.82e}$$

Finally, we solve the above set of differential equations, with the obvious 'boundary' conditions

$$\breve{\alpha}(K) = 0,\ \ \bar{\alpha}_A(K) = 0,\ \ \bar{\alpha}_S(K) = 0,\ \ \bar{Q}(K_{\max}) = 1,\ \ \bar{Q}_p(K_{\max}) = 1, \tag{4.83}$$

and take the limit $k \to K_e$ with $K_e \ll K$. Solving first for

$$\breve{\alpha}(k) = \frac{8\pi^2}{3}\frac{Ro^2 D_1\theta_K}{E_\nu\left(E_\nu + E_\eta\right)}\left[\frac{1}{4}\left(\frac{1}{k^4} - \frac{1}{K^4}\right) - \frac{\left(3E_\nu + E_\eta\right)\Gamma\tilde{\omega}}{12E_\nu\left(E_\nu + E_\eta\right)}\left(\frac{1}{k^6} - \frac{1}{K^6}\right)\right], \tag{4.84}$$

and substituting into (4.82c), we write down the final expressions for the coefficients describing corrections from fluctuations to the means

$$\bar{Q} = 1 - \frac{4\pi^2}{45}\frac{Ro^2}{E_\nu^2 E_\eta}\frac{D_0}{K_e^4}\left[1 - \frac{3\left(E_\nu + 2E_\eta\right)}{8E_\nu E_\eta}\frac{\Gamma\tilde{\omega}}{K_e^2}\right], \tag{4.85a}$$

$$\bar{Q}_p = 1 + \frac{4\pi^2}{45}\frac{Ro^2(E_\nu - 4E_\eta)}{E_\nu^2 E_\eta (E_\nu + E_\eta)}\frac{D_0}{K_e^4}\left[1 - \frac{3}{8}\frac{E_\nu^3 + 4E_\nu^2 E_\eta - 15E_\nu E_\eta^2 - 8E_\eta^3}{E_\nu E_\eta (E_\nu + E_\eta)(E_\nu - 4E_\eta)}\frac{\Gamma\tilde{\varpi}}{K_e^2}\right], \quad (4.85b)$$

$$\begin{aligned}\bar{\alpha}_A = & \frac{2\pi^2}{3}\frac{Ro^2 D_1}{E_\nu(E_\nu + E_\eta)K_e^4}\left\{1 - \frac{(3E_\nu + E_\eta)\Gamma\tilde{\varpi}}{3E_\nu(E_\nu + E_\eta)K_e^2}\right. \\ & - \frac{\pi^2}{3}\frac{Ro^2 D_0}{E_\nu(E_\nu + E_\eta)^2 K_e^4} + \frac{2\pi^2}{45}\frac{Ro^2(21E_\nu + 5E_\eta)D_0\Gamma\tilde{\varpi}}{E_\nu^2(E_\nu + E_\eta)^3 K_e^6} \\ & \left. - \frac{\pi^2}{27}\frac{Ro^2(5E_\nu + E_\eta)(3E_\nu + E_\eta)D_0(\Gamma\tilde{\varpi})^2}{E_\nu^3(E_\nu + E_\eta)^4 K_e^8}\right\},\end{aligned} \quad (4.85c)$$

$$\bar{\alpha}_S = \frac{4\pi^2}{15}\frac{Ro^2(Ro\tilde{\beta})^2 H}{E_\nu^2 E_\eta (E_\nu + E_\eta)}\frac{D_1}{K_e^6}\left[1 - \frac{3(E_\nu^2 + 5E_\nu E_\eta + 2E_\eta^2)}{8E_\nu E_\eta (E_\nu + E_\eta)}\frac{\Gamma\tilde{\varpi}}{K_e^2}\right]. \quad (4.85d)$$

The transformation (4.44) can be used if one needs to get back to the dimensional variables, as done later in section 4.4.

4.3.2 Role of diffusivities

It is very important to understand the role of viscosity and magnetic diffusivity in the adopted approach. First of all we recall, that the assumption

$$Ro\tilde{\beta} \ll 1, \quad (4.86)$$

is utilized to facilitate the Fourier integrals which need to be computed at each step of the renormalization procedure by expanding the integrands in Taylor series in $Ro\tilde{\beta}$, more precisely the γ_u^{-1} factors as in (4.59). One must realize, this implies we assume (cf (4.43))

$$|-\mathrm{i}\omega + E_\nu k^2| \gg \frac{H(Ro\tilde{\beta})^2(\mathbf{k}\cdot\langle\mathbf{B}\rangle)^2}{|-\mathrm{i}\omega + E_\eta k^2|}, \quad (4.87)$$

for all admissible values of ω, thus in particular for $\omega \to 0$. Consequently, we require

$$E_\nu E_\eta K_e^2 \gg H(Ro\tilde{\beta})^2\left(\frac{\mathbf{K}_e}{K_e}\cdot\langle\mathbf{B}\rangle\right)^2 = \mathcal{O}\left((Ro\tilde{\beta})^2\right), \quad (4.88)$$

where $K_e = 2\pi\ell/\ell_e = \mathcal{O}(1)$ is the non-dimensional wave number, hence the approach is valid only when both the diffusivities are non-zero. It follows that the astrophysically relevant limit of low diffusivities $E_\nu \ll 1$ and/or $E_\eta \ll 1$ must be taken with care, incorporating (4.88). In other words setting $\nu = 0$ makes the problem ill-posed, because the ω-integrals cease to converge. The case of $\eta = 0$ is similar and at least some of the ω-integrals diverge in this limit. Hence we must require

$$\nu \neq 0, \quad \text{and} \quad \eta \neq 0, \tag{4.89}$$

for soundness and consistency of the mathematical approach. Moreover, we can easily see from the analysis in section 4.2, that for weak and stationary turbulence setting $\eta = 0$ leads to vanishing of the α-effect for all values of $Ro\tilde{\beta}$, despite irregularity of the ω-integral due to the perturbation expansion of the integrand for $Ro\tilde{\beta} \ll 1$. It is clear from equation (4.6) (or from (3.154a and b) and $\Delta(\omega, \omega') \sim \delta(\omega + \omega')$ in stationary case, since the integrand for $\eta = 0$ is an odd function of ω thus the entire integral over $-\infty < \omega < \infty$ must vanish, even without expanding in Taylor series in $Ro\tilde{\beta}$). Consequently, we arrive at the well-known result, that in a weak turbulence the mean field dynamo does not operate when the magnetic diffusivity is too weak and negligible (cf [21]). However, the case of non-stationary turbulence is different, as then the mean field dynamo operates even in the absence of fluid's resistivity. Equation (4.88) involves non-dimensional wave number K_e and $\langle \mathbf{B} \rangle$, which by definition are order unity quantities. It follows, that the validity of the current approach can be expressed in a more straightforward manner by the Reynolds numbers $Re = \mathcal{U}\ell/\nu$ and $Rm = \mathcal{U}\ell/\eta$ in the following way

$$\frac{\nu\eta}{\mathcal{U}^2\ell^2} = \frac{1}{ReRm} \gg \tilde{\beta}^2, \quad \text{or equivalently} \quad \frac{Pm}{Rm_L^2\varepsilon^2} \gg \tilde{\beta}^2, \tag{4.90}$$

where $Rm_L = \mathcal{U}L/\eta$ is the magnetic Reynolds number based on the global length scale L, a useful parameter in the description of the MHD turbulence in natural systems, in particular the theory of natural dynamos. The latter restriction is fully consistent with the adopted weakly nonlinear approach, i.e. the assumption that the evolution of turbulent fluctuations is weakly nonlinear. From the point of view of applications to real astrophysical systems the condition (4.90) is quite restrictive, since the astrophysical length scales are very large and the condition (4.90) demands that the magnetic Reynolds number Rm_L which is proportional to L is 'not too large'. Nevertheless the upper bound for Rm_L can still be large, in particular for high-Pm systems.

4.4 Mean field dynamics

The renormalized mean-field equations take the dimensional form (cf (4.75a and b))

$$\frac{\partial\langle \mathbf{U}\rangle}{\partial t} + (\langle \mathbf{U}\rangle \cdot \nabla)\langle \mathbf{U}\rangle = -\nabla\left(\frac{\langle p\rangle}{\rho} + \bar{Q}_p\frac{\langle \mathbf{B}\rangle^2}{2}\right) + \bar{Q}(\langle \mathbf{B}\rangle \cdot \nabla)\langle \mathbf{B}\rangle + \nu\nabla^2\langle \mathbf{U}\rangle, \tag{4.91a}$$

$$\frac{\partial\langle \mathbf{B}\rangle}{\partial t} = \nabla \times (\langle \mathbf{U}\rangle \times \langle \mathbf{B}\rangle) + (\bar{\alpha}_A - \bar{\alpha}_S\langle B\rangle^2)\nabla \times \langle \mathbf{B}\rangle - \bar{\alpha}_S\nabla\langle B\rangle^2 \times \langle \mathbf{B}\rangle + \eta\nabla^2\langle \mathbf{B}\rangle, \tag{4.91b}$$

where the coefficients $\bar{\alpha}_A$, $\bar{\alpha}_S$, $\bar{Q}$ and $\bar{Q}_p$ include the effect of the turbulent fluctuations on the means. The coefficient $\bar{\alpha}_A$ can give the effect of amplification of the mean magnetic field energy, whereas $\bar{\alpha}_S$ results from action of the Lorentz force and is responsible for saturation. The two coefficients $\bar{Q}$ and $\bar{Q}_P$ in the mean-field Navier–Stokes equation describe the renormalized Lorentz force. The general non-dimensional formulae for all the renormalized coefficients $\bar{\alpha}_A$, $\bar{\alpha}_S$, $\bar{Q}$ and $\bar{Q}_P$ are provided in (4.85a–d). In dimensional form the mean α-effect is described by

$$\bar{\alpha}_A = \frac{2\pi^2}{3}\frac{D_1}{\nu(\nu+\eta)K_e^4}\left\{1 - \frac{(3\nu+\eta)\Gamma\tilde{\omega}}{3\nu(\nu+\eta)K_e^2} - \frac{\pi^2 D_0}{3\nu(\nu+\eta)^2 K_e^4} + \frac{\pi^2(3\nu+\eta)D_0\Gamma\tilde{\omega}}{9\nu^2(\nu+\eta)^3 K_e^6}\left[\frac{2(21\nu+5\eta)}{5(3\nu+\eta)} - \frac{(5\nu+\eta)\Gamma\tilde{\omega}}{3\nu(\nu+\eta)K_e^2}\right]\right\}, \tag{4.92a}$$

$$\bar{\alpha}_S = \frac{4\pi^2}{15}\frac{D_1}{\nu^2\eta(\nu+\eta)K_e^6}\left[1 - \frac{3(\nu^2+5\nu\eta+2\eta^2)}{8\nu\eta(\nu+\eta)}\frac{\Gamma\tilde{\omega}}{K_e^2}\right]. \tag{4.92b}$$

The above formulae were obtained within the limit $Ro \ll Ro\tilde{\beta} \ll 1$ and $\tilde{\omega}t \ll 1$, $\tilde{\omega} \ll \nu K_e^2, \eta K_e^2$, which were put forward in order to make analytical progress and obtain the final formulae which include both the effect of non-stationarity and the 'real turbulence' effect, i.e. the influence of nonlinear evolution of the turbulent fluctuations. The non-equilibrium corrections are proportional to the slow kinetic energy production rate, since by (4.17) we have

$$\Gamma\tilde{\omega} \approx \frac{\nu K_e^2}{2\pi^2 D_0}\frac{\partial\langle e_k\rangle}{\partial t}, \tag{4.93}$$

and we recall that $D_0 > 0$. Naturally, from the physical point of view the assumption of weak non-stationarity (weak departure from equilibrium, $\tilde{\omega} \ll \nu K_e^2, \eta K_e^2$) is not necessary and in many applications rather unrealistic. In fact, the departure from equilibrium, measured by the frequency shift $\tilde{\omega}$, could be stronger than diffusion, especially in high-conductivity plasma, if the viscosity is also not too large (as assumed e.g. in section 3.2; see also (4.13) and the comment above explaining that a regularizing assumption of $D_0 = D_0(\omega) \sim \omega^2$ and $D_1 = D_1(\omega) \sim \omega^2$ at $\omega \to 0$ allows us to remove the constraint of non-zero magnetic diffusivity, and then the α-effect consists solely of the non-equilibrium contributions which remain non-zero even when $\eta = 0$). However, within the limit considered in the current chapter the non-equilibrium corrections in the above expressions (4.92a and b) are small by assumption (in contrast to the section 3.4 where the mean magnetic field was assumed strong and the γ_u factors were not expanded in powers of $Ro\tilde{\beta}$).

Before we proceed to the discussion of the obtained results, we first associate the forcing parameters D_0 and D_1 with the leading-order form of the turbulent kinetic energy $\langle u'^2\rangle$ and the turbulent kinetic helicity $\langle \mathbf{u}' \cdot \mathbf{w}'\rangle$, respectively. At the leading order the velocity and magnetic fields are given by (3.152) and thus the vorticity takes the form

$$\hat{w}_i(\mathbf{q}) = \mathrm{i}\epsilon_{ijk}k_j\hat{u}_k(\mathbf{q}) \approx \frac{\mathrm{i}}{\gamma_\nu(\mathbf{q})}\epsilon_{ijk}k_j\hat{f}_k(\mathbf{q}). \tag{4.94}$$

The kinetic energy and helicity of the turbulent fluctuations, with the aid of (4.45) can be calculated as follows

$$\langle e_k \rangle = \frac{1}{2}\langle u_i'(\mathbf{x}, t)u_i'(\mathbf{x}, t)\rangle \approx \frac{1}{2}\int d^4q \int d^4q' \frac{e^{i[(\mathbf{k}+\mathbf{k}')\cdot\mathbf{x}-(\omega+\omega')t]}}{\gamma_\nu(\mathbf{q}')\gamma_\nu(\mathbf{q})}\langle \hat{f}_i(\mathbf{q})\hat{f}_i(\mathbf{q}')\rangle$$
$$\approx \frac{4\pi^2 D_0}{\nu}\int_{K_e}^{K_\nu} \frac{dk}{k^3}\left(1 + \Gamma\tilde{\omega}t - \frac{\Gamma\tilde{\omega}}{2\nu k^2}\right) \approx \frac{2\pi^2 D_0}{\nu K_e^2}\left(1 - \frac{\Gamma\tilde{\omega}}{4\nu K_e^2}\right), \tag{4.95}$$

$$\langle h_k \rangle = \langle u_i'(\mathbf{x}, t)w_i'(\mathbf{x}, t)\rangle \approx i\epsilon_{ijk}\int d^4q \int d^4q' \frac{k_j' e^{i[(\mathbf{k}+\mathbf{k}')\cdot\mathbf{x}-(\omega+\omega')t]}}{\gamma_\nu(\mathbf{q})\gamma_\nu(\mathbf{q}')}\langle \hat{f}_i(\mathbf{q})\hat{f}_k(\mathbf{q}')\rangle$$
$$\approx -\frac{8\pi^2 D_1}{\nu}\int_{K_e}^{K_\nu} \frac{dk}{k^3}\left(1 + \Gamma\tilde{\omega}t - \frac{\Gamma\tilde{\omega}}{2\nu k^2}\right) \approx -\frac{4\pi^2 D_1}{\nu K_e^2}\left(1 - \frac{\Gamma\tilde{\omega}}{4\nu K_e^2}\right), \tag{4.96}$$

which yields

$$D_0 \approx \frac{\nu K_e^2}{2\pi^2}\left(1 + \frac{\Gamma\tilde{\omega}}{4\nu K_e^2}\right)\langle e_k \rangle, \qquad D_1 \approx -\frac{\nu K_e^2}{4\pi^2}\left(1 + \frac{\Gamma\tilde{\omega}}{4\nu K_e^2}\right)\langle h_k \rangle, \tag{4.97}$$

and by (4.93)

$$\Gamma\tilde{\omega} \approx \frac{\partial\langle e_k \rangle}{\partial t}. \tag{4.98}$$

Consequently,

$$\bar{\alpha}_A \approx -\frac{\langle h_k \rangle}{6(\nu + \eta)K_e^2}\left[1 - \frac{9\nu + \eta}{12(\nu + \eta)\nu K_e^2}\frac{\partial\langle e_k \rangle}{\partial t} - \frac{\langle e_k \rangle}{6(\nu + \eta)^2 K_e^2}\right], \tag{4.99a}$$

$$\bar{\alpha}_S \approx -\frac{\langle h_k \rangle}{15\nu\eta(\nu + \eta)K_e^4}\left[1 - \frac{3\nu^2 + 13\nu\eta + 4\eta^2}{8\eta(\nu + \eta)\nu K_e^2}\frac{\partial\langle e_k \rangle}{\partial t}\right]. \tag{4.99b}$$

The following general conclusions can be drawn. Firstly, when $\Gamma\tilde{\omega}>0$ ($\partial_t\langle e_k \rangle > 0$), that is when the mean kinetic energy increases with time (cf (4.95) and also (4.46)) the nonlinear effect proportional to the turbulent energy enhances the non-equilibrium effect, but both have opposite sign to the standard diffusive α-effect. On the other hand for $\Gamma\tilde{\omega}<0$ ($\partial_t\langle e_k \rangle < 0$), i.e. when the kinetic energy decreases with time, the non-equilibrium effect enhances the diffusive one, but the nonlinearity tends to quench them. Naturally, the entire α-effect vanishes when turbulence is reflectionally symmetric and $\langle h_k \rangle = 0$.

The magnitude of the α-effect could be roughly estimated by adopting the following estimate of the turbulent helicity, $\langle h_k \rangle \approx \mathcal{U}^2/\ell_e$, where $\mathcal{U}^2 \approx 2\langle e_k \rangle$; in such a case

$$|\,\bar{\alpha}_A\,| \approx \frac{\mathcal{U}^2}{12\pi(\nu + \eta)K_e}\left[1 - \frac{9\nu + \eta}{12(\nu + \eta)}\frac{\Gamma\tilde{\omega}}{\nu K_e^2} - \frac{\mathcal{U}^2}{12(\nu + \eta)^2 K_e^2}\right], \tag{4.100}$$

but we still need some physical interpretation of the non-stationarity factor $\Gamma\tilde{\omega}$, that is, the energy production rate. We can get this by adopting the physical picture based on argumentation put forward at the end of section 3.3.1. Turbulence in natural systems can rarely be considered statistically stationary and very often is forced at a few distinct frequencies. For example, rapidly rotating satellites of massive planets or giant planets orbiting near a massive star (so called 'hot Jupiters') experience strong tidal and Coriolis forces. Due to the effect of tidal synchronization of the orbital and rotational motion, these two forcings typically act on close frequencies; it follows, that the small difference between these frequencies could determine the frequency shift $\tilde{\omega} = \Omega_{\text{orbital}} - \Omega_{\text{rot}}$ in the turbulence spectral correlation tensor[2]

$$\left\langle \hat{u}_i(\mathbf{k}, \omega)\hat{u}_j(\mathbf{k}', \omega')\right\rangle = \left[\mathcal{D}_0(k, \omega)P_{ij}(\mathbf{k}) + \mathrm{i}\mathcal{D}_1(k, \omega)\epsilon_{ijk}k_k\right]\delta(\mathbf{k} + \mathbf{k}') \times \left[\delta(\omega + \omega') + \frac{\Gamma}{2\mathrm{i}}(\delta(\omega + \omega' + \tilde{\omega}) - \delta(\omega + \omega' - \tilde{\omega}))\right]. \tag{4.101}$$

On the other hand, in the case of ISM turbulence, as argued in section 3.4, a sensible measure of departure from equilibrium is provided by the difference between the ejection rates for type I and type II supernovae, which Korpi *et al* [22] estimate at $\Gamma\tilde{\omega}{\approx}10^{-10}\,/s$. In many systems, however, the turbulence is obviously statistically non-stationary, but two distinct driving frequencies are not evident.

4.4.1 Force-free mode, $\langle \mathbf{J}\rangle \propto \langle \mathbf{B}\rangle$

As in the previous chapter, in order to complete the dynamical picture we study the energy evolution of the force-free mode in the mean magnetic field defined by

$$\nabla \times \langle \mathbf{B}\rangle = \kappa\langle \mathbf{B}\rangle, \quad \kappa = \mathcal{O}(\varepsilon), \quad \kappa \neq 0, \tag{4.102}$$

so that the Lorentz force from the mean field vanishes, $\langle \mathbf{j}\rangle \times \langle \mathbf{B}\rangle = 0$, in other words $(\langle \mathbf{B}\rangle \cdot \nabla)\langle \mathbf{B}\rangle$ becomes potential. We also set the mean velocity to zero

$$\langle \mathbf{U}\rangle = 0, \tag{4.103}$$

and then the magnetic energy density $E_m = \langle B\rangle^2/2\mu_0\rho$ of a force-free mode is governed by the equation

$$\frac{\partial E_m}{\partial t} = 2\kappa(\bar{\alpha}_A - \kappa\eta - 2\bar{\alpha}_S E_m)E_m. \tag{4.104}$$

The full analytic solution is easily accessible and can be expressed in the form

[2] Recall, that such a spectral correlation tensor leads to sinusoidal time dependence of the velocity correlations as in (4.46). This is why $\tilde{\omega}$ has the interpretation of a small difference between two driving frequencies, since for $\mathbf{f}_1 = \hat{\mathbf{f}}_1 e^{\mathrm{i}\Omega_1 t} + c.\,c.$, and $\mathbf{f}_2 = \hat{\mathbf{f}}_2 e^{\mathrm{i}\Omega_2 t} + c.\,c.$ one obtains $\langle \mathbf{f}_1\mathbf{f}_2\rangle \sim \sin((\Omega_1 - \Omega_2)t)$, if the frequency difference $\Omega_1 - \Omega_2$ is small enough and survives the averaging (of course by the relation $\hat{u}_i(\mathbf{q}) \approx \hat{f}_i(\mathbf{q})/\gamma_\nu$ we expect a similar correlation tensor for the velocity and the forcing).

$$E_m(t) = \frac{E_\infty \exp[2\kappa(\bar{\alpha}_A - \kappa\eta)t]}{\exp[2\kappa(\bar{\alpha}_A - \kappa\eta)t] - 1 + \dfrac{E_\infty}{E_0}}, \tag{4.105}$$

where

$$E_\infty = \frac{\bar{\alpha}_A - \kappa\eta}{2\bar{\alpha}_S}, \tag{4.106}$$

is the asymptotic energy value at $t \to \infty$ in the case when the magnetic energy does not decay, i.e. for

$$\frac{\bar{\alpha}_A}{\kappa} > \eta, \tag{4.107}$$

which then also implies $\bar{\alpha}_S\kappa > 0$, so that E_∞ in (4.106) is positive. In such a case three distinct regimes are present for $E_\infty/E_0 \neq 1, t > 0$ (for $E_\infty/E_0 = 1$ the magnetic energy remains unaltered for all times)

$$\frac{E_\infty}{E_0} < 1 \quad \Longrightarrow \quad (\dot{E}_m(t) < 0,\ \ddot{E}_m(t) > 0), \tag{4.108a}$$

$$1 < \frac{E_\infty}{E_0} \leqslant 2 \quad \Longrightarrow \quad (\dot{E}_m(t) > 0,\ \ddot{E}_m(t) < 0), \tag{4.108b}$$

$$\frac{E_\infty}{E_0} > 2 \quad \Longrightarrow \quad (\dot{E}_m(t) > 0,\ \ddot{E}_m(t < t_I) > 0,\ \ddot{E}_m(t > t_I) < 0), \tag{4.108c}$$

where the upper dot denotes the time derivative and since E_∞ is a function of system parameters (4.106), the conditions on E_∞/E_0 on the left hand side should be regarded as conditions on the initial magnetic energy value relative to the ratio $(\bar{\alpha}_A - \kappa\eta)/2\bar{\alpha}_S$, with the α-coefficients given in (4.92a and b). Furthermore, in the above t_I marks the inflexion point of $E_m(t)$ defined by $\ddot{E}_m(t_I) = 0$,

$$t_I = \frac{\ln\left(\dfrac{E_\infty}{E_0} - 1\right)}{4\kappa\bar{\alpha}_S E_\infty}, \quad \text{when} \quad \frac{E_\infty}{E_0} > 2. \tag{4.109}$$

figure 4.1 depicts the magnetic energy evolution for the two cases of (4.108b) and (4.108c), corresponding to energy growth. The period of exponential amplification (regime with EMF linear in $\langle \mathbf{B} \rangle$) belongs to the third case (4.108c), present for $0 < t < t_I$. It follows from (4.109), that according to intuition, the larger the ratio E_∞/E_0 the longer the period of exponential amplification. Let us now consider two limiting cases of particular interest in geo- and astrophysics, when the magnetic Prandtl number $Pm = \nu/\eta$ is small, as in the case of the Earth's core, and large as in the ISM. This will allow us to gain a better picture of the dynamical effect of statistical non-stationarity in the dynamo process. At small Pm we get from (4.92a and b)

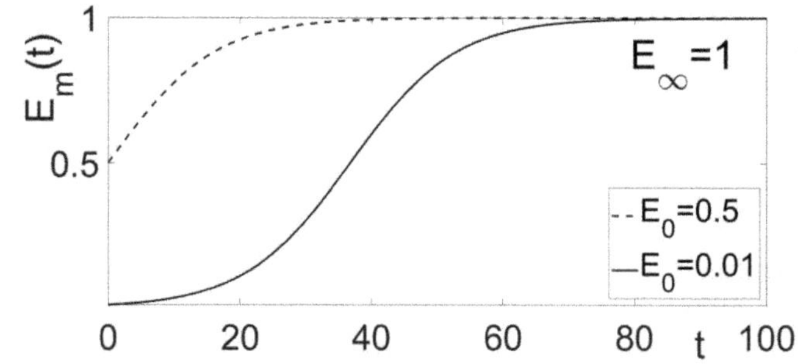

Figure 4.1. Saturation of the magnetic energy of a force-free mode for two cases: $E_\infty/E_0 = 100$ (solid line) corresponding to (4.108c) and $E_\infty/E_0 = 2$ corresponding to (4.108b) (dashed line); in both cases $\kappa\tilde{\alpha}_A - \kappa^2\eta = 1/16$, the asymptotic energy value is $E_\infty = 1$, but the initial energies differ. This implies $\kappa\tilde{\alpha}_S = 1/32$ for both cases.

$$\bar{\alpha}_A \approx \frac{2\pi^2}{3}\frac{D_1}{\nu\eta K_e^4}\left[1 - \frac{1}{3}\frac{\Gamma\tilde{\omega}}{\nu K_e^2} - \frac{\pi^2 D_0}{3\nu\eta^2 K_e^4} + \frac{2\pi^2 D_0\Gamma\tilde{\omega}}{9\nu^2\eta^2 K_e^6}\left(1 - \frac{\Gamma\tilde{\omega}}{6\nu K_e^2}\right)\right], \tag{4.110a}$$

$$\bar{\alpha}_S \approx \frac{4\pi^2}{15}\frac{D_1}{\nu^2\eta^2 K_e^6}\left(1 - \frac{3}{4}\frac{\Gamma\tilde{\omega}}{\nu K_e^2}\right), \tag{4.110b}$$

and when Pm is large

$$\bar{\alpha}_A = \frac{2\pi^2}{3}\frac{D_1}{\nu^2 K_e^4}\left[1 - \frac{\Gamma\tilde{\omega}}{\nu K_e^2} - \frac{\pi^2 D_0}{3\nu^3 K_e^4} + \frac{\pi^2 D_0\Gamma\tilde{\omega}}{3\nu^4 K_e^6}\left(\frac{14}{5} - \frac{5\Gamma\tilde{\omega}}{3\nu K_e^2}\right)\right], \tag{4.111a}$$

$$\bar{\alpha}_S = \frac{4\pi^2}{15}\frac{D_1}{\nu^3\eta K_e^6}\left(1 - \frac{3}{8}Pm\frac{\Gamma\tilde{\omega}}{\nu K_e^2}\right). \tag{4.111b}$$

The saturation energy (4.106) is then approximately

$$E_\infty \approx \frac{5}{4}\nu\eta K_e^2\left[1 + \frac{5}{12}\left(1 - \frac{6}{5}\frac{12\pi\eta^2 K_e^2}{L\,|\,\langle h_k\rangle\,|}\right)\frac{\Gamma\tilde{\omega}}{\nu K_e^2} - \frac{\langle e_k\rangle}{6\eta^2 K_e^2} - \frac{12\pi\eta^2 K_e^2}{L\,|\,\langle h_k\rangle\,|}\right], \tag{4.112}$$

at $Pm \ll 1$ and

$$E_\infty \approx \frac{5}{4}\nu\eta K_e^2\left[1 + \frac{3}{8}\left(1 - \frac{12\pi\nu\eta K_e^2}{L\,|\,\langle h_k\rangle\,|}\right)Pm\frac{\Gamma\tilde{\omega}}{\nu K_e^2} - \frac{\langle e_k\rangle}{6\nu^2 K_e^2} - \frac{12\pi\nu\eta K_e^2}{L\,|\,\langle h_k\rangle\,|}\right], \tag{4.113}$$

at $Pm \gg 1$, when the following conditions resulting from (4.107) are satisfied

$$1 - \frac{12\pi\eta^2 K_e^2}{L\,|\,\langle h_k\rangle\,|} > 0, \quad \text{at } Pm \ll 1, \tag{4.114}$$

$$1 - \frac{12\pi\nu\eta K_e^2}{L \mid \langle h_k \rangle \mid} > 0, \quad \text{at } Pm \gg 1. \tag{4.115}$$

Clearly, the non-equilibrium effect is more strongly manifested when the magnetic diffusivity is weak, $Pm \gg 1$, since then the correction proportional to $\Gamma\tilde{\omega}$ is greater. Moreover, in the case when the turbulent kinetic energy increases with time, $\Gamma\tilde{\omega}>0$, the non-equilibrium effects tend to increase the final saturation energy of the mean magnetic field E_∞, but when $\Gamma\tilde{\omega}<0$, i.e. the turbulent kinetic energy decreases with time, the saturation energy of the mean magnetic field is quenched (strictly speaking at small Pm this is true when the kinetic helicity is strong enough, so that the term in the parentheses in (4.112) next to $\Gamma\tilde{\omega}/\nu K_e^2$ is positive, which is not guaranteed by (4.114)). This is an interesting result in light of the conclusions drawn from (4.99a and b)—see discussion below. When $\Gamma\tilde{\omega}<0$ both the coefficients $\bar{\alpha}_A$ and $\bar{\alpha}_S$ are enhanced by non-stationarity, but since the first is responsible for amplification and the second for saturation of the magnetic energy, which in the final state depends on the ratio $\bar{\alpha}_A/\bar{\alpha}_S$ (cf (4.106)), the field saturates at a smaller energy than the equilibrium value, because the $\bar{\alpha}_S$ coefficient is influenced by the non-equilibrium effect more strongly. On the other hand, when $\Gamma\tilde{\omega}>0$ the quenching effect of non-stationarity on the $\bar{\alpha}_S$ coefficient is also stronger than on the $\bar{\alpha}_A$ coefficient and the resulting saturation energy E_∞ is larger than in equilibrium. This shows that the non-equilibrium dynamo effect is highly non-trivial and requires caution.

Finally, it is of interest to point out that within the kept order of accuracy the value of the coefficient $\bar{\alpha}_S$ in (4.92b), associated with the value of saturation energy, is unaltered with respect to that obtained from the linearized equations (cf (4.9)). This is in contrast to the value of the coefficient $\bar{\alpha}_A$, which is clearly affected by the weakly nonlinear evolution of the turbulent fluctuations. It would require inclusion of terms up to $\mathcal{O}(Ro(Ro\tilde{\beta})^5)$ and $\mathcal{O}(Ro^3(Ro\tilde{\beta})^3)$ in our asymptotic analysis in order to obtain a weakly nonlinear turbulent correction to the coefficient $\bar{\alpha}_S$.

4.5 Summary of main results

The weakly nonlinear renormalization procedure for non-stationary stirred MHD turbulence has allowed us to obtain the renormalized set of the mean field MHD equations, including the complete form of the effective mean EMF and of the large-scale Lorentz force in the mean Navier–Stokes equation; the theory was developed within the limit of a small 'Rossby' number $Ro \ll 1$ defined as a relative measure of the fluid's inertia with respect to the stirring force. With the use of the final expressions for the $\bar{\alpha}_A$ and $\bar{\alpha}_S$ coefficients (4.99a and b) we were able to estimate the final value at which the energy of a force-free mode of the mean magnetic field saturates and concluded that the non-equilibrium effects play a more significant role when the magnetic diffusivity is weak, $\eta \ll \nu$. The non-equilibrium corrections to the α-effect turned out proportional to the slow kinetic energy production rate $\Gamma\tilde{\omega}\approx\partial\langle e_k\rangle/\partial t$, cf (4.99a and b). Furthermore, when the non-equilibrium turbulence experiences growth of the turbulent kinetic energy, $\Gamma\tilde{\omega}>0$, the $\bar{\alpha}_A$, $\bar{\alpha}_S$ coefficients are quenched by non-stationarity, but because the saturational effect, $\bar{\alpha}_S$, is more significantly weakened, the final energy of saturation of the mean magnetic field is typically larger than in the case of equilibrium turbulence. The opposite is true, when the turbulent kinetic energy decreases in time, $\Gamma\tilde{\omega}<0$, i.e. although both the coefficients $\bar{\alpha}_A$, $\bar{\alpha}_S$ are enhanced by the non-equilibrium effects, the final energy of saturation is smaller than in equilibrium, because the $\bar{\alpha}_S$ coefficient is more strongly affected. The nonlinear effects in turbulence cooperate with the non-equilibrium effects when, $\Gamma\tilde{\omega}>0$ but compete in the opposite case. Let us also recall that the measure of non-stationarity $\tilde{\omega}$ has the physical interpretation of the beating frequency, i.e. small frequency difference between interacting distinct waves in the turbulent wave field.

Of particular astrophysical interest is the limit of extremely low and negligible resistivity, when the diffusively controlled mechanism, based on resistive phase shift between the kinetic and magnetic components of waves is no longer effective, i.e. $\bar{\alpha}_{st} = 0$. *This limit corresponds to the ISM and plasma in AGN (active galactic nuclei). In the case of such systems, the non-equilibrium dynamo effect which operates regardless of weak resistivity may prove to be a crucial and perhaps dominant factor for their hydromagnetic dynamo action. The* α*-effect induced by non-stationarity of turbulence when* $\eta = 0$ *is defined by equation (4.13), i.e. it is proportional to the energy production rate* $\tilde{\omega}$.

References

[1] Tobias S M, Cattaneo F and Boldyrev S 2013 MHD dynamos and turbulence *Ten Chapters in Turbulence* ed P A Davidson, Y Kaneda and K R Sreenivasan (Cambridge: Cambridge University Press) 351–404 pp

[2] Tobias S M 2021 The turbulent dynamo *J. Fluid Mech.* **912** P1

[3] Ma S and Mazenko G F 1975 Critical dynamics of ferromagnets in 6-ϵ dimensions: general discussion and detailed calculation *Phys. Rev.* B **11** 4077–100

[4] Wyld H W Jr 1961 Formulation of the theory of turbulence in an incompressible fluid *Ann. Phys.* **14** 143–65

[5] Forster D, Nelson D R and Stephen M J 1977 Large-distance and long-time properties of a randomly stirred fluid *Phys. Rev.* A **16** 732–49

[6] Moffatt H K 1981 Some developments in the theory of turbulence *J. Fluid Mech.* **106** 27–47

[7] Moffatt H K 1983 Transport effects associated with turbulence with particular attention to the influence of helicity *Rep. Prog. Phys.* **46** 621–64

[8] Yakhot V and Orszag S A 1986 Renormalization group analysis of turbulence. I. Basic theory *J. Sci. Comp.* **1** 3–51

[9] McComb W D and Watt A G 1990 Conditional averaging procedure for the elimination of the small-scale modes from incompressible fluid turbulence at high Reynolds numbers *Phys. Rev. Lett.* **65** 3281–4

[10] McComb W D and Watt A G 1992 Two-field theory of incompressible-fluid turbulence *Phys. Rev.* A **46** 4797–812

[11] McComb W D, Roberts W and Watt A G 1992 Conditional-averaging procedure for problems with mode-mode coupling *Phys. Rev.* A **45** 3507–15

[12] Smith L M and Reynolds W C 1992 On the Yakhot-Orszag renormalization group method for deriving turbulence statistics and models *Phys. Fluids* A **4** 364–90
[13] Lam S H 1992 On the RNG theory of turbulence *Phys. Fluids* A **4** 1007–17
[14] Eyink G L 1994 The renormalization group method in statistical hydrodynamics *Phys. Fluids* **6** 3063–78
[15] Smith L M and Woodruff S L 1998 Renormalization-group analysis of turbulence *Annu. Rev. Fluid Mech.* **30** 275–310
[16] McComb W D 2014 *Homogeneous, Isotropic Turbulence. Phenomenology, Renormalization and Statistical Closures* (Oxford: Oxford University Press)
[17] Kleeorin N and Rogachevskii I 1994 Effective Ampère force in developed magnetohydrodynamic turbulence *Phys. Rev.* E **50** 2716–30
[18] Mizerski K A 2021 Renormalization group analysis of the magnetohydrodynamic turbulence and dynamo *J. Fluid Mech.* **926** A13
[19] Mizerski K A 2020 Renormalization group analysis of the turbulent hydromagnetic dynamo: effect of non-stationarity *Astrophys. J. Suppl. Ser.* **251** 21
[20] Moffatt H K 1972 An approach to a dynamic theory of dynamo action in a rotating conducting fluid *J. Fluid Mech.* **53** 385–99
[21] Moffatt H K and Dormy E 2019 *Self-exciting Fluid Dynamos* (Cambridge: Cambridge University Press)
[22] Korpi M J, Brandenburg A, Shukurov A, Tuominen I and Nordlund Å 1999 A supernova-regulated interstellar medium: simulations of the turbulent multiphase medium *Astrophys. J.* **514** L99–102

Chapter 5

Non-equilibrium turbulent dynamo from joint action and evolutional history of cross- and kinetic helicities

In this chapter we will remove the effect of the stirring force in the turbulent flow, i.e. we will set $\mathbf{f} = 0$ in equation (2.6a–c) and study the effect of non-stationarity on the turbulent hydromagnetic dynamo process with the use of the Two-Scale Direct-Interaction Approximation (TSDIA). This method allows us to remove the stirring force, but instead we need to assume some statistical properties of the background turbulence. Nevertheless, this approach allows us to explicitly relate the mean turbulent Reynolds and Maxwell stresses and the mean electromotive force (EMF) to the spectral characteristics of turbulence, such as, e.g., the mean energy, as well as kinetic and cross-helicity, through consideration of the Green's response functions, which describe the responses of the turbulent flow and magnetic field to infinitesimal perturbations, cf e.g. [1–6]. It will be demonstrated, that the non-equilibrium effects can enhance the dynamo process through creation of a yet additional correction to the α-effect, unexplored in previous chapters, when the magnetohydrodynamic (MHD) turbulence is both helical and cross-helical. This effect, identified and thoroughly described in [7], is based on the turbulent infinitesimal-impulse *cross*-response functions, which do not affect turbulent flows in equilibrium. We repeat here the analysis of [7], providing some more details on the mathematical approach and discuss the evolution equation of the cross-helicity (2.55), its sources and sinks in MHD turbulence, hence the possibility of a coexistence of the kinetic and cross-helicities.

In the previous chapters we showed that the non-equilibrium α-effect is proportional to the energy production rate resulting from presence of a statistically non-stationary stirring force and it is oscillatory on time-scales induced by the forcing, which could be long. Here we will utilize the TSDIA approach, in order to extract the effect of non-stirred, non-equilibrium turbulence on the large-scale

doi:10.1088/978-0-7503-6362-4ch5

hydromagnetic dynamo. In other words, in order to study the dynamo process in an incompressible conducting fluid we will consider the set of dynamical equations (2.6a–c) with $\mathbf{f} = 0$ and some additional assumptions about statistical properties of the background turbulence. It will be demonstrated that in non-equilibrium turbulence the quantity $\langle \mathbf{u}' \cdot \mathbf{j}' \rangle$ plays a significant role in generation of the large-scale EMF through the α-effect and the effect of $\langle \mathbf{u}' \cdot \mathbf{j}' \rangle$ vanishes in stationary turbulence.

5.1 Outline of the TSDIA approach

TSDIA is based on the idea of introduction of the Green's functions of turbulence (which dates back to Kraichnan 1959, 1965 [8, 9]), that is, tensorial functions describing responses of the velocity and magnetic fields to infinitesimal-impulse-force perturbations and a perturbative treatment of weak large-scale inhomogeneities, through an ad hoc introduction of two scales of turbulence in space and time, related by the same small parameter of expansion. The Green's functions include self- and cross-interaction response functions between the turbulent fields $\mathbf{u}'(\mathbf{x}, t)$ and $\mathbf{b}'(\mathbf{x}, t)$ and their infinitesimal perturbations. In this section we will present the outline of the TSDIA formulation with the final resulting expression for the turbulent EMF and for more details on the TSDIA approach the reader is referred to textbooks such as Yoshizawa (1998) [3] and Yokoi (2020) [10].

Let us introduce a small parameter δ[1] and define slow and fast spatial and temporal variables

$$\boldsymbol{\xi} = \mathbf{x}, \qquad \mathbf{X} = \delta\mathbf{x}, \qquad \tau = t, \qquad T = \delta t, \tag{5.1}$$

so that

$$\frac{\partial}{\partial t} = \frac{\partial}{\partial \tau} + \delta\frac{\partial}{\partial T}, \qquad \nabla = \nabla_{\xi} + \delta\nabla_{\mathbf{X}}. \tag{5.2}$$

By assumption, the large-scale fields depend only on the slow variables, $\langle \mathbf{U}\rangle(\mathbf{X}, T)$ and the fluctuations depend on both, $\mathbf{u}'(\boldsymbol{\xi}, \mathbf{X}; \tau, T)$. We also define the Fourier transform, involving Galilean transformation to the frame moving with the velocity $\langle \mathbf{U}\rangle$

$$u_i'(\boldsymbol{\xi}, \mathbf{X}; \tau, T) = \int \mathrm{d}^3k \hat{u}_i(\mathbf{k}, \mathbf{X}; \tau, T) \mathrm{e}^{-\mathrm{i}\mathbf{k}\cdot(\boldsymbol{\xi} - \langle \mathbf{U}\rangle\tau)}, \tag{5.3a}$$

$$b_i'(\boldsymbol{\xi}, \mathbf{X}; \tau, T) = \int \mathrm{d}^3k \hat{b}_i(\mathbf{k}, \mathbf{X}; \tau, T) \mathrm{e}^{-\mathrm{i}\mathbf{k}\cdot(\boldsymbol{\xi} - \langle \mathbf{U}\rangle\tau)}, \tag{5.3b}$$

but the explicit dependence on the slow variables $\mathbf{X}$ and T will be typically suppressed in notation for clarity, hence the Fourier transforms will be denoted by $\hat{u}_i(\mathbf{k}; \tau)$, $\hat{b}_i(\mathbf{k}; \tau)$. The system of two-scale differential equations (2.13a–c) in the wave number space with $\mathbf{f} = 0$ is written as

[1] We note that δ is essentially the same as ε in chapter 4, although here δ denotes scale separation, both spatial and temporal; for historical reasons in this chapter we keep this notation the same as in [3].

$$\frac{\partial \hat{u}_i(\mathbf{k};\tau)}{\partial \tau} + \nu k^2 \hat{u}_i(\mathbf{k};\tau) + \mathrm{i}k_j \langle B \rangle_j \hat{b}_i(\mathbf{k};\tau)$$
$$- \mathrm{i}M_{ij\ell}(\mathbf{k}) \int \mathrm{d}^3p \int \mathrm{d}^3q\, \delta(\mathbf{k}-\mathbf{p}-\mathbf{q}) \left[\hat{u}_j(\mathbf{p};\tau)\hat{u}_\ell(\mathbf{q};\tau) - \hat{b}_j(\mathbf{p};\tau)\hat{b}_\ell(\mathbf{q};\tau) \right]$$
$$= \delta \left[-P_{ij}(\mathbf{k}) \frac{\hat{D}\hat{u}_j(\mathbf{k};\tau)}{DT_{\mathrm{I}}} - P_{ij}(\mathbf{k})\hat{u}_m(\mathbf{k};\tau) \left(\frac{\partial \langle U \rangle_j}{\partial X_m} + \epsilon_{mj\ell}\Omega_{0\ell} \right) \right. \tag{5.4}$$
$$\left. + \langle B \rangle_j \frac{\partial \hat{b}_i(\mathbf{k};\tau)}{\partial X_{j\mathrm{I}}} + P_{ij}(\mathbf{k})\hat{b}_m(\mathbf{k};\tau) \frac{\partial \langle B \rangle_j}{\partial X_m} \right],$$

$$-\mathrm{i}k_j \hat{u}_j(\mathbf{k};\tau) + \delta \frac{\partial \hat{u}_j(\mathbf{k};\tau)}{\partial X_{j\mathrm{I}}} = 0, \tag{5.5}$$

$$\frac{\partial \hat{b}_i(\mathbf{k};\tau)}{\partial \tau} + \eta k^2 \hat{b}_i(\mathbf{k};\tau) + \mathrm{i}k_j \langle B \rangle_j \hat{u}_i(\mathbf{k};\tau)$$
$$+ \mathrm{i}N_{ij\ell}(\mathbf{k}) \int \mathrm{d}^3p \int \mathrm{d}^3q\, \delta(\mathbf{k}-\mathbf{p}-\mathbf{q}) \left[\hat{b}_j(\mathbf{p};\tau)\hat{u}_\ell(\mathbf{q};\tau) - \hat{u}_j(\mathbf{p};\tau)\hat{b}_\ell(\mathbf{q};\tau) \right]$$
$$= \delta \left[-P_{ij}(\mathbf{k}) \frac{\hat{D}\hat{b}_j(\mathbf{k};\tau)}{DT_{\mathrm{I}}} + P_{ij}(\mathbf{k})\hat{b}_m(\mathbf{k};\tau) \left(\frac{\partial \langle U \rangle_j}{\partial X_m} + \epsilon_{mj\ell}\Omega_{0\ell} \right) \right. \tag{5.6}$$
$$\left. + \langle B \rangle_j \frac{\partial \hat{u}_i(\mathbf{k};\tau)}{\partial X_{j\mathrm{I}}} - P_{ij}(\mathbf{k})\hat{u}_m(\mathbf{k};\tau) \frac{\partial \langle B \rangle_j}{\partial X_m} \right],$$

$$-\mathrm{i}k_j \hat{b}_j(\mathbf{k};\tau) + \delta \frac{\partial \hat{b}_j(\mathbf{k};\tau)}{\partial X_{j\mathrm{I}}} = 0, \tag{5.7}$$

where

$$\left(\nabla_{\mathbf{X}\mathrm{I}}, \frac{D}{DT_{\mathrm{I}}} \right) = \exp(-\mathrm{i}\mathbf{k}\cdot\langle \mathbf{U} \rangle \tau) \left(\nabla_{\mathbf{X}}, \frac{D}{DT} \right) \exp(\mathrm{i}\mathbf{k}\cdot\langle \mathbf{U} \rangle \tau), \tag{5.8}$$

and

$$M_{ijk}(\mathbf{k}) = \frac{1}{2}\left[k_j P_{ik}(\mathbf{k}) + k_k P_{ij}(\mathbf{k}) \right], \qquad N_{ijk}(\mathbf{k}) = \frac{1}{2}\left(k_j \delta_{ik} - k_k \delta_{ij} \right), \tag{5.9}$$

$$P_{ij}(\mathbf{k}) = \delta_{ij} - \frac{k_i k_j}{k^2}. \tag{5.10}$$

Within the TSDIA procedure, which involves renormalization (see section 5.2 below), the operators M_{ijk} and N_{ijk} correspond to point vertices showing the wave-number conservation among the nonlinear mode coupling with $\delta(\mathbf{k}-\mathbf{p}-\mathbf{q})$. Moreover, the terms

$$\delta\nabla_X \cdot (\mathbf{u}'\mathbf{u}' - \mathbf{b}'\mathbf{b}' - \langle\mathbf{u}'\mathbf{u}'\rangle + \langle\mathbf{b}'\mathbf{b}'\rangle), \text{ and } \delta\nabla_X \times (\mathbf{u}' \times \mathbf{b}' - \langle\mathbf{u}' \times \mathbf{b}'\rangle), \tag{5.11}$$

in the Navier–Stokes (5.4) and induction (5.6) equations, respectively, have been neglected, but note that the neglected terms are fully at the order δ, and the influence of nonlinearities $(\mathbf{u}' \cdot \nabla)\mathbf{u}'$, $(\mathbf{b}' \cdot \nabla)\mathbf{b}'$, $(\mathbf{u}' \cdot \nabla)\mathbf{b}'$ and $(\mathbf{b}' \cdot \nabla)\mathbf{u}'$ is included at the leading order $\mathcal{O}(1)$; therefore the TSDIA approximation is by no means related to the first-order smoothing approximation since in the latter the nonlinear terms in the equations for fluctuations are entirely neglected.

In (5.4) and (5.6) we have also introduced the 'co-rotational' derivative

$$\frac{\hat{D}\hat{u}_i}{DT} = \frac{\partial\hat{u}_i}{\partial T} + \langle U\rangle_j\frac{\partial\hat{u}_i}{\partial X_j} + \epsilon_{jik}\Omega_{0k}\hat{u}_j, \tag{5.12}$$

with

$$\boldsymbol{\Omega}_0 = \boldsymbol{\Omega}/\delta = \mathcal{O}(1), \tag{5.13}$$

and we assume, that the background rotation is of comparable magnitude with the mean velocity gradient, $\boldsymbol{\Omega} = \mathcal{O}(\delta)$. The co-rotational derivative possesses the useful property of being invariant with respect to transformations between the inertial frame of reference and the frame rotating with angular velocity $\boldsymbol{\Omega}$, contrary to the advective derivative

$$\frac{D\hat{u}_i}{DT} = \frac{\partial\hat{u}_i}{\partial T} + \langle U\rangle_j\frac{\partial\hat{u}_i}{\partial X_j}, \tag{5.14}$$

which is not.

The next step is to expand the dynamical fields such as $\hat{u}(\mathbf{k}; \tau)$ and $\hat{b}(\mathbf{k}; \tau)$ with respect to the scale parameter δ, and utilize the solenoidal conditions (5.5) and (5.7) to split the fields $\hat{u}(\mathbf{k}; \tau)$ and $\hat{b}(\mathbf{k}; \tau)$ into two components: along the wave vector $\mathbf{k}$ and orthogonal to $\mathbf{k}$,

$$\hat{u}_i(\mathbf{k}; \tau) = \sum_{n=0}^{\infty}\delta^n\hat{u}_{i\,n}(\mathbf{k}; \tau) - \sum_{n=0}^{\infty}\delta^{n+1}\mathrm{i}\frac{k_i}{k^2}\frac{\partial}{\partial X_{j\mathrm{I}}}\hat{u}_{j\,n}(\mathbf{k}; \tau), \tag{5.15}$$

where

$$\forall_n \; k_i\hat{u}_{i\,n}(\mathbf{k}; \tau) = 0. \tag{5.16}$$

Furthermore, in the TSDIA formalism it is assumed that all the effects which generate inhomogeneity and anisotropy of fluctuations are weak, therefore, not only are the frame rotation $\boldsymbol{\Omega}$ and the large-scale shear $\nabla\langle\mathbf{U}\rangle$ assumed weak, but the mean magnetic field $\langle\mathbf{B}\rangle$ as well. Hence, in addition to the δ-expansion, each of the $\hat{u}_{i\,n}(\mathbf{k}; \tau)$ terms is also expanded with respect to the weak mean field in the following way

$$\hat{u}_i(\mathbf{k}; \tau) = \sum_{n=0}^{\infty}\sum_{m=0}^{\infty}\delta^n\hat{u}_{i\,nm}(\mathbf{k}; \tau) - \sum_{n=0}^{\infty}\sum_{m=0}^{\infty}\delta^{n+1}\mathrm{i}\frac{k_i}{k^2}\frac{\partial}{\partial X_{j\mathrm{I}}}\hat{u}_{j\,nm}(\mathbf{k}; \tau). \tag{5.17}$$

It follows that the lowest-order fields $\hat{u}_{i\,00}$ correspond to the homogeneous and isotropic turbulence, uninfluenced by $\langle\mathbf{B}\rangle$, $\boldsymbol{\Omega}$, and $\nabla\langle\mathbf{U}\rangle$, and inhomogeneities and anisotropies enter at higher orders in δ ($n \geqslant 1$) and in the mean field $\langle\mathbf{B}\rangle$ ($m \geqslant 1$). Using the expansions (5.17) for the velocity $\hat{u}(\mathbf{k};\,\tau)$ and analogously for the magnetic field $\hat{b}(\mathbf{k};\,\tau)$, we write down equations (5.4) and (5.6) at the first three orders, i.e.

$$\begin{aligned}&\frac{\partial\hat{u}_{i\,00}(\mathbf{k};\,\tau)}{\partial\tau} + \nu k^2\hat{u}_{i\,00}(\mathbf{k};\,\tau)\\&- \mathrm{i}M_{ij\ell}(\mathbf{k})\int \mathrm{d}^3p\int \mathrm{d}^3q\;\delta(\mathbf{k}-\mathbf{p}-\mathbf{q})\Big[\hat{u}_{j\,00}(\mathbf{p};\,\tau)\hat{u}_{\ell\,00}(\mathbf{q};\,\tau)\\&\qquad -\hat{b}_{j\,00}(\mathbf{p};\,\tau)\hat{b}_{\ell\,00}(\mathbf{q};\,\tau)\Big] = 0,\end{aligned} \tag{5.18a}$$

$$\begin{aligned}&\frac{\partial\hat{b}_{i}(\mathbf{k};\,\tau)}{\partial\tau} + \eta k^2\hat{b}_{i}(\mathbf{k};\,\tau)\\&+ \mathrm{i}N_{ij\ell}(\mathbf{k})\int \mathrm{d}^3p\int \mathrm{d}^3q\;\delta(\mathbf{k}-\mathbf{p}-\mathbf{q})\Big[\hat{b}_{j\,00}(\mathbf{p};\,\tau)\hat{u}_{\ell\,00}(\mathbf{q};\,\tau)\\&\qquad -\hat{u}_{j\,00}(\mathbf{p};\,\tau)\hat{b}_{\ell\,00}(\mathbf{q};\,\tau)\Big] = 0,\end{aligned} \tag{5.18b}$$

for $n = m = 0$, and

$$\begin{aligned}&\frac{\partial\hat{u}_{i\,01}(\mathbf{k};\,\tau)}{\partial\tau} + \nu k^2\hat{u}_{i\,01}(\mathbf{k};\,\tau)\\&- 2\mathrm{i}M_{ij\ell}(\mathbf{k})\int \mathrm{d}^3p\int \mathrm{d}^3q\;\delta(\mathbf{k}-\mathbf{p}-\mathbf{q})\big[\hat{u}_{j\,00}(\mathbf{p};\,\tau)\hat{u}_{\ell\,01}(\mathbf{q};\,\tau)\\&\qquad -\hat{b}_{j\,00}(\mathbf{p};\,\tau)\hat{b}_{\ell\,01}(\mathbf{q};\,\tau)\big] = -\mathrm{i}k_j\langle B\rangle_j\hat{b}_{i\,00}(\mathbf{k};\,\tau),\end{aligned} \tag{5.19a}$$

$$\begin{aligned}&\frac{\partial\hat{b}_{i\,01}(\mathbf{k};\,\tau)}{\partial\tau} + \eta k^2\hat{b}_{i\,01}(\mathbf{k};\,\tau)\\&+ 2\mathrm{i}N_{ij\ell}(\mathbf{k})\int \mathrm{d}^3p\int \mathrm{d}^3q\;\delta(\mathbf{k}-\mathbf{p}-\mathbf{q})\big[\hat{b}_{j\,00}(\mathbf{p};\,\tau)\hat{u}_{\ell\,01}(\mathbf{q};\,\tau)\\&\qquad -\hat{u}_{j\,00}(\mathbf{p};\,\tau)\hat{b}_{\ell\,01}(\mathbf{q};\,\tau)\big] = -\mathrm{i}k_j\langle B\rangle_j\hat{u}_{i\,00}(\mathbf{k};\,\tau),\end{aligned} \tag{5.19b}$$

for $n = 0$, $m = 1$ (containing the leading order effect of the mean magnetic field $\langle\mathbf{B}\rangle$, still in absence of $\boldsymbol{\Omega}$, and $\nabla\langle\mathbf{U}\rangle$), and finally

$$\begin{aligned}&\frac{\partial\hat{u}_{i\,10}(\mathbf{k};\,\tau)}{\partial\tau} + \nu k^2\hat{u}_{i\,10}(\mathbf{k};\,\tau)\\&- 2\mathrm{i}M_{ij\ell}(\mathbf{k})\int \mathrm{d}^3p\int \mathrm{d}^3q\;\delta(\mathbf{k}-\mathbf{p}-\mathbf{q})\big[\hat{u}_{j\,00}(\mathbf{p};\,\tau)\hat{u}_{\ell\,10}(\mathbf{q};\,\tau) - \hat{b}_{j\,00}(\mathbf{p};\,\tau)\hat{b}_{\ell\,10}(\mathbf{q};\,\tau)\big]\\&\quad = -P_{ij}(\mathbf{k})\frac{\hat{D}\hat{u}_{j\,00}(\mathbf{k};\,\tau)}{DT_{\mathrm{I}}} - P_{ij}(\mathbf{k})\hat{u}_{m\,00}(\mathbf{k};\,\tau)\left(\frac{\partial\langle U\rangle_j}{\partial X_m} + \epsilon_{mj\ell}\Omega_{0\ell}\right)\\&\qquad + \langle B\rangle_j\frac{\partial\hat{b}_{i\,00}(\mathbf{k};\,\tau)}{\partial X_{j\mathrm{I}}} + P_{ij}(\mathbf{k})\hat{b}_{m\,00}(\mathbf{k};\,\tau)\frac{\partial\langle B\rangle_j}{\partial X_m},\end{aligned} \tag{5.20a}$$

$$\frac{\partial \hat{b}_{i\,10}(\mathbf{k};\tau)}{\partial\tau} + \eta k^2 \hat{b}_{i\,10}(\mathbf{k};\tau)$$
$$+ 2\mathrm{i}N_{ij\ell}(\mathbf{k}) \int \mathrm{d}^3p \int \mathrm{d}^3q\ \delta(\mathbf{k}-\mathbf{p}-\mathbf{q})\Big[\hat{b}_{j\,00}(\mathbf{p};\tau)\hat{u}_{\ell\,10}(\mathbf{q};\tau) - \hat{u}_{j\,00}(\mathbf{p};\tau)\hat{b}_{\ell\,10}(\mathbf{q};\tau)\Big]$$
$$= -P_{ij}(\mathbf{k})\frac{\hat{D}\hat{b}_{j\,00}(\mathbf{k};\tau)}{DT_{\mathrm{I}}} + P_{ij}(\mathbf{k})\hat{b}_{m\,00}(\mathbf{k};\tau)\left(\frac{\partial\langle U\rangle_j}{\partial X_m} + \epsilon_{mj\ell}\Omega_{0\ell}\right) \tag{5.20b}$$
$$+ \langle B\rangle_j \frac{\partial \hat{u}_{i\,00}(\mathbf{k};\tau)}{\partial X_{j\mathrm{I}}} - P_{ij}(\mathbf{k})\hat{u}_{m\,00}(\mathbf{k};\tau)\frac{\partial\langle B\rangle_j}{\partial X_m},$$

for $n = 1$, $m = 0$ (containing the leading order effect of all three anisotropy factors $\langle\mathbf{B}\rangle$, $\mathbf{\Omega}$, $\nabla\langle\mathbf{U}\rangle$). The right-hand sides of (5.19a and b) and (5.20a and b) can be regarded as source terms for the evolution of $\hat{u}_{01}(\mathbf{k};\tau)$, $\hat{b}_{01}(\mathbf{k};\tau)$ and $\hat{u}_{10}(\mathbf{k};\tau)$, $\hat{b}_{10}(\mathbf{k};\tau)$, respectively. Note that in (5.20a) and (5.20b) the contributions from the second terms in (5.17), i.e. from

$$-\delta\mathrm{i}\frac{k_i}{k^2}\frac{\partial}{\partial X_{j\mathrm{I}}}\hat{u}_{j\,00}(\mathbf{k};\tau), \text{ and } -\delta\mathrm{i}\frac{k_i}{k^2}\frac{\partial}{\partial X_{j\mathrm{I}}}\hat{b}_{j\,00}(\mathbf{k};\tau), \tag{5.21}$$

have been neglected, since eventually in calculation of the mean EMF, $\epsilon_{ijk}\langle u_j' b_k'\rangle$, they produce third-order correlations only, of the type $\left\langle \hat{u}_{i\,00}(\mathbf{p};\tau)\hat{u}_{j\,00}(\mathbf{q};\tau)\hat{u}_{k\,00}(\mathbf{r};\tau)\right\rangle$ (cf section 5.2 below) and these are neglected within the TSDIA approach due to assumed Gaussianity of the background turbulence, given by the lowest-order fields $\hat{\mathbf{u}}_{00}$, $\hat{\mathbf{b}}_{00}$; for more details see [1, 3].

5.1.1 The Green's response functions

The direct-interaction approximation (DIA) is based on introduction of tensorial spectral Green's functions associated with the 0th order fields, i.e. equations (5.18a and b). The responses of the turbulent fields to infinitesimal-impulse-force disturbances describe the mutual interactions between the velocity and magnetic fields. Namely, four independent Green's functions are introduced, G_{uu} which represents the response of the velocity field $\mathbf{u}$ to the velocity perturbation $\mathbf{u}$, G_{ub} which describes the response of $\mathbf{u}$ to the magnetic perturbation $\mathbf{b}$, G_{bu} the response of $\mathbf{b}$ to the velocity perturbation $\mathbf{u}$, and G_{bb} the response of magnetic field $\mathbf{b}$ to the magnetic perturbation $\mathbf{b}$. They are defined by the following evolution equations

$$\frac{\partial G_{ij}^{(uu)}(\mathbf{k};\tau,\tau')}{\partial\tau} + \nu k^2 G_{ij}^{(uu)}(\mathbf{k};\tau,\tau')$$
$$- 2\mathrm{i}M_{ikm}(\mathbf{k}) \int \mathrm{d}^3p \int \mathrm{d}^3q\ \delta(\mathbf{k}-\mathbf{p}-\mathbf{q})\Big[\hat{u}_{k\,00}(\mathbf{p};\tau)G_{mj}^{(uu)}(\mathbf{q};\tau,\tau') \tag{5.22a}$$
$$- \hat{b}_{k\,00}(\mathbf{p};\tau)G_{mj}^{(bu)}(\mathbf{q};\tau,\tau')\Big] = P_{ij}(\mathbf{k})\delta(\tau-\tau'),$$

$$\frac{\partial G_{ij}^{(bb)}(\mathbf{k};\,\tau,\,\tau')}{\partial\tau}+\eta k^2 G_{ij}^{(bb)}(\mathbf{k};\,\tau,\,\tau')$$
$$+\,2\mathrm{i}N_{ikm}(\mathbf{k})\int \mathrm{d}^3p\int \mathrm{d}^3q\;\delta(\mathbf{k}-\mathbf{p}-\mathbf{q})\Big[\hat{b}_{k\,00}(\mathbf{p};\,\tau)G_{mj}^{(ub)}(\mathbf{q};\,\tau,\,\tau')$$
$$-\,\hat{u}_{k\,00}(\mathbf{p};\,\tau)G_{mj}^{(bb)}(\mathbf{q};\,\tau,\,\tau')\Big]=P_{ij}(\mathbf{k})\delta(\tau-\tau'),\tag{5.22b}$$

$$\frac{\partial G_{ij}^{(ub)}(\mathbf{k};\,\tau,\,\tau')}{\partial\tau}+\nu k^2 G_{ij}^{(ub)}(\mathbf{k};\,\tau,\,\tau')$$
$$-\,2\mathrm{i}M_{ikm}(\mathbf{k})\int \mathrm{d}^3p\int \mathrm{d}^3q\;\delta(\mathbf{k}-\mathbf{p}-\mathbf{q})\Big[\hat{u}_{k\,00}(\mathbf{p};\,\tau)G_{mj}^{(ub)}(\mathbf{q};\,\tau,\,\tau')$$
$$-\hat{b}_{k\,00}(\mathbf{p};\,\tau)G_{mj}^{(bb)}(\mathbf{q};\,\tau,\,\tau')\Big]=0,\tag{5.22c}$$

$$\frac{\partial G_{ij}^{(bu)}(\mathbf{k};\,\tau,\,\tau')}{\partial\tau}+\eta k^2 G_{ij}^{(bu)}(\mathbf{k};\,\tau,\,\tau')$$
$$+\,2\mathrm{i}N_{ikm}(\mathbf{k})\int \mathrm{d}^3p\int \mathrm{d}^3q\;\delta(\mathbf{k}-\mathbf{p}-\mathbf{q})\Big[\hat{b}_{k\,00}(\mathbf{p};\,\tau)G_{mj}^{(uu)}(\mathbf{q};\,\tau,\,\tau')$$
$$-\hat{u}_{k\,00}(\mathbf{p};\,\tau)G_{mj}^{(bu)}(\mathbf{q};\,\tau,\,\tau')\Big]=0.\tag{5.22d}$$

The 01 and 10 fields in (5.19a and b) and (5.20a and b) are now easily expressed with the use of the Green's functions in the following way

$$\hat{u}_{i\,01}(\mathbf{k};\,\tau)=-\,\mathrm{i}\langle B\rangle_m k_m\int_{-\infty}^{\tau}\mathrm{d}\tau' G_{ij}^{(uu)}(\mathbf{k};\,\tau,\,\tau')\hat{b}_{j\,00}(\mathbf{k};\,\tau')$$
$$-\,\mathrm{i}\langle B\rangle_m k_m\int_{-\infty}^{\tau}\mathrm{d}\tau' G_{ij}^{(ub)}(\mathbf{k};\,\tau,\,\tau')\hat{u}_{j\,00}(\mathbf{k};\,\tau'),\tag{5.23a}$$

$$\hat{b}_{i\,01}(\mathbf{k};\,\tau)=-\,\mathrm{i}\langle B\rangle_m k_m\int_{-\infty}^{\tau}\mathrm{d}\tau' G_{ij}^{(bu)}(\mathbf{k};\,\tau,\,\tau')\hat{b}_{j\,00}(\mathbf{k};\,\tau')$$
$$-\,\mathrm{i}\langle B\rangle_m k_m\int_{-\infty}^{\tau}\mathrm{d}\tau' G_{ij}^{(bb)}(\mathbf{k};\,\tau,\,\tau')\hat{u}_{j\,00}(\mathbf{k};\,\tau'),\tag{5.23b}$$

$$\hat{u}_{i\,10}(\mathbf{k};\,\tau)=-\,P_{jn}(\mathbf{k})\int_{-\infty}^{\tau}\mathrm{d}\tau' G_{ij}^{(uu)}(\mathbf{k};\,\tau,\,\tau')\frac{\hat{D}\hat{u}_{n\,00}(\mathbf{k};\,\tau')}{DT_{\mathrm{I}}}$$
$$-\,P_{jn}(\mathbf{k})\int_{-\infty}^{\tau}\mathrm{d}\tau' G_{ij}^{(ub)}(\mathbf{k};\,\tau,\,\tau')\frac{\hat{D}\hat{b}_{n\,00}(\mathbf{k};\,\tau')}{DT_{\mathrm{I}}}$$
$$+\,P_{jn}(\mathbf{k})\frac{\partial\langle B\rangle_n}{\partial X_m}\int_{-\infty}^{\tau}\mathrm{d}\tau'\Big[G_{ij}^{(uu)}(\mathbf{k};\,\tau,\,\tau')\hat{b}_{m\,00}(\mathbf{k};\,\tau')$$
$$-\,G_{ij}^{(ub)}(\mathbf{k};\,\tau,\,\tau')\hat{u}_{m\,00}(\mathbf{k};\,\tau')\Big]\tag{5.23c}$$
$$-\,P_{jn}(\mathbf{k})\left(\frac{\partial\langle U\rangle_n}{\partial X_m}+\epsilon_{mn\ell}\Omega_{0\ell}\right)\int_{-\infty}^{\tau}\mathrm{d}\tau'\Big[G_{ij}^{(uu)}(\mathbf{k};\,\tau,\,\tau')\hat{u}_{m\,00}(\mathbf{k};\,\tau')$$
$$-G_{ij}^{(ub)}(\mathbf{k};\,\tau,\,\tau')\hat{b}_{m\,00}(\mathbf{k};\,\tau')\Big]$$
$$+\,\langle B\rangle_m\int_{-\infty}^{\tau}\mathrm{d}\tau'\left[G_{ij}^{(ub)}(\mathbf{k};\tau,\tau')\frac{\partial\hat{u}_{j\,00}(\mathbf{k};\,\tau')}{\partial X_{m\mathrm{I}}}+G_{ij}^{(uu)}(\mathbf{k};\,\tau,\,\tau')\frac{\partial\hat{b}_{j\,00}(\mathbf{k};\,\tau')}{\partial X_{m\mathrm{I}}}\right],$$

$$
\begin{aligned}
\hat{b}_{i\,10}(\mathbf{k};\,\tau) = &- P_{jn}(\mathbf{k})\int_{-\infty}^{\tau} \mathrm{d}\tau' G_{ij}^{(bu)}(\mathbf{k};\,\tau,\,\tau')\frac{\hat{D}\hat{u}_{n\,00}(\mathbf{k};\,\tau')}{DT_{\mathrm{I}}} \\
&- P_{jn}(\mathbf{k})\int_{-\infty}^{\tau} \mathrm{d}\tau' G_{ij}^{(bb)}(\mathbf{k};\,\tau,\,\tau')\frac{\hat{D}\hat{b}_{n\,00}(\mathbf{k};\,\tau')}{DT_{\mathrm{I}}} \\
&+ P_{jn}(\mathbf{k})\frac{\partial\langle B\rangle_n}{\partial X_m}\int_{-\infty}^{\tau} \mathrm{d}\tau'\Big[G_{ij}^{(bu)}(\mathbf{k};\,\tau,\,\tau')\hat{b}_{m\,00}(\mathbf{k};\,\tau') \\
&\qquad - G_{ij}^{(bb)}(\mathbf{k};\,\tau,\,\tau')\hat{u}_{m\,00}(\mathbf{k};\,\tau')\Big] \\
&- P_{jn}(\mathbf{k})\left(\frac{\partial\langle U\rangle_n}{\partial X_m} + \epsilon_{mn\ell}\Omega_{0\ell}\right)\int_{-\infty}^{\tau} \mathrm{d}\tau'\Big[G_{ij}^{(bu)}(\mathbf{k};\,\tau,\,\tau')\hat{u}_{m\,00}(\mathbf{k};\,\tau') \\
&\qquad - G_{ij}^{(bb)}(\mathbf{k};\,\tau,\,\tau')\hat{b}_{m\,00}(\mathbf{k};\,\tau')\Big] \\
&+ \langle B\rangle_m\int_{-\infty}^{\tau} \mathrm{d}\tau'\left[G_{ij}^{(bb)}(\mathbf{k};\tau,\tau')\frac{\partial\hat{u}_{j\,00}(\mathbf{k};\tau')}{\partial X_{m\mathrm{I}}} + G_{ij}^{(bu)}(\mathbf{k};\tau,\tau')\frac{\partial\hat{b}_{j\,00}(\mathbf{k};\,\tau')}{\partial X_{m\mathrm{I}}}\right],
\end{aligned}
\tag{5.23d}
$$

which is a great advantage of the TSDIA approach.

5.1.2 Statistical properties of the homogeneous and isotropic background turbulence

We assume that the statistics of lowest-order fields (indexed with 00) is given. In other words the physical picture we have in mind is such, that the properties of the background turbulence are known, and inhomogeneities and anisotropy are introduced on top of the strong turbulent $\mathbf{u}_{00}$ and $\mathbf{b}_{00}$ fields at higher orders in δ and the mean magnetic field. The background turbulence is uninfluenced by the large-scale field and rotation, hence is assumed homogeneous and isotropic, and defined by the following correlation functions

$$
\left\langle \hat{f}_{i\,00}(\mathbf{k};\,\tau)\hat{g}_{j\,00}(\mathbf{k}';\,\tau')\right\rangle = \left[P_{ij}(\mathbf{k})Q_{fg}(k;\,\tau,\,\tau') + \frac{1}{2}\mathrm{i}\epsilon_{ijk}\frac{k_k}{k^2}H_{fg}(k;\,\tau,\,\tau') \right]\delta(\mathbf{k}+\mathbf{k}'), \tag{5.24}
$$

where $\mathbf{f}_{00}$ and $\mathbf{g}_{00}$ represent one of $\mathbf{u}_{00}$ and $\mathbf{b}_{00}$, and the subscripts f and g correspond to u and b accordingly.

The formalism of the DIA introduced by Kraichnan [8, 9] involves a renormalization procedure, which effectively allows one to treat the propagators G_{uu}, G_{bb}, G_{ub}, and G_{bu} as deterministic variables, i.e. we write

$$
\langle G_{ij}^{(uu)}(\mathbf{k};\,\tau,\,\tau')\rangle = G_{ij}^{(uu)}(\mathbf{k};\,\tau,\,\tau') = P_{ij}(\mathbf{k})G^{(uu)}(k;\,\tau,\,\tau'), \tag{5.25a}
$$

$$
\langle G_{ij}^{(bb)}(\mathbf{k};\,\tau,\,\tau')\rangle = G_{ij}^{(bb)}(\mathbf{k};\,\tau,\,\tau') = P_{ij}(\mathbf{k})G^{(bb)}(k;\,\tau,\,\tau'), \tag{5.25b}
$$

$$
\langle G_{ij}^{(ub)}(\mathbf{k};\,\tau,\,\tau')\rangle = G_{ij}^{(ub)}(\mathbf{k};\,\tau,\,\tau') = P_{ij}(\mathbf{k})G^{(ub)}(k;\,\tau,\,\tau'), \tag{5.25c}
$$

$$
\langle G_{ij}^{(bu)}(\mathbf{k};\,\tau,\,\tau')\rangle = G_{ij}^{(bu)}(\mathbf{k};\,\tau,\,\tau') = P_{ij}(\mathbf{k})G^{(bu)}(k;\,\tau,\,\tau'); \tag{5.25d}
$$

for detailed treatment see [3]. It is also useful at this stage to write down explicitly the relations between the spectral functions, Q_{uu}, Q_{bb}, Q_{ub}, H_{uu}, H_{bb}, H_{ub}, and H_{bu} and the turbulent correlation functions such as the turbulent kinetic energy, magnetic energy, cross-helicity, kinetic helicity and current helicity associated with the background turbulence, i.e. the lowest order fields $\mathbf{u}_{00}$, $\mathbf{b}_{00}$. We start with the quantity $\left\langle \mathbf{u}'_{00} \cdot \mathbf{j}'_{00} \right\rangle$ which will play an important role in the theory of non-equilibrium α-effect, developed below,

$$\begin{aligned}\left\langle \mathbf{u}'_{00}(\mathbf{x}, \tau) \cdot \mathbf{j}'_{00}(\mathbf{x}, \tau') \right\rangle &= -\mathrm{i}\epsilon_{ijk} \int \mathrm{d}k \int \mathrm{d}k' k'_j \langle \hat{u}'_{00i}(\mathbf{k}; \tau) \hat{b}'_{00k}(\mathbf{k}'; \tau') \rangle \mathrm{e}^{-\mathrm{i}(\mathbf{k}+\mathbf{k}')\cdot \mathbf{x}} \\ &= \int \mathrm{d}k H_{ub}(k; \tau, \tau') = \int \mathrm{d}k H_{bu}(k; \tau', \tau);\end{aligned} \tag{5.26}$$

moreover,

$$\int \mathrm{d}^3k Q_{uu}(k; \tau, \tau) = \frac{1}{2} \langle \mathbf{u}'^2_{00} \rangle, \quad \int \mathrm{d}^3k Q_{bb}(k; \tau, \tau) = \frac{1}{2} \langle \mathbf{b}'^2_{00} \rangle, \tag{5.27}$$

$$\int \mathrm{d}^3k Q_{ub}(k; \tau, \tau) = \langle \mathbf{u}'_{00} \cdot \mathbf{b}'_{00} \rangle, \tag{5.28}$$

$$\int \mathrm{d}^3k H_{uu}(k; \tau, \tau) = \langle \mathbf{u}'_{00} \cdot \mathbf{w}'_{00} \rangle, \quad \int \mathrm{d}^3k H_{bb}(k; \tau, \tau) = \left\langle \mathbf{b}'_{00} \cdot \mathbf{j}'_{00} \right\rangle. \tag{5.29}$$

We also note, that under the current assumptions of isotropy and homogeneity of the background turbulence when $\nu = \eta$ we have

$$G_{ij}^{(ub)} = G_{ij}^{(bu)}, \qquad G_{ij}^{(uu)} = G_{ij}^{(bb)} \tag{5.30}$$

provided that we also require

$$\begin{aligned}k_m P_{ik}(\mathbf{k}) \int \mathrm{d}^3p \int \mathrm{d}^3q \; \delta(\mathbf{k} - \mathbf{p} - \mathbf{q}) \; &\left[\hat{u}_{k\,00}(\mathbf{p}; \tau) G_{mj}^{(uu)}(\mathbf{q}; \tau, \tau') \right. \\ &\left. - \hat{b}_{k\,00}(\mathbf{p}; \tau) G_{mj}^{(ub)}(\mathbf{q}; \tau, \tau') \right] = 0,\end{aligned} \tag{5.31a}$$

$$\begin{aligned}k_m P_{ik}(\mathbf{k}) \int \mathrm{d}^3p \int \mathrm{d}^3q \; \delta(\mathbf{k} - \mathbf{p} - \mathbf{q}) \; &\left[\hat{b}_{k\,00}(\mathbf{p}; \tau) G_{mj}^{(uu)}(\mathbf{q}; \tau, \tau') \right. \\ &\left. - \hat{u}_{k\,00}(\mathbf{p}; \tau) G_{mj}^{(ub)}(\mathbf{q}; \tau, \tau') \right] = 0,\end{aligned} \tag{5.31b}$$

to be satisfied; the latter equations are trivially satisfied by solutions of (5.22a–d) in the form (5.25a–d). Moreover, when the diffusivities are unequal but weak

$$\nu \ll \Omega \ell^2, \qquad \eta \ll \Omega \ell^2, \tag{5.32}$$

and the additional requirements (5.31a and b) are imposed, the relation

$$G_{ij}^{(ub)} \approx G_{ij}^{(bu)}, \tag{5.33}$$

still holds.

5.2 The large-scale electromotive force and the α-effect

We can now express the turbulent EMF

$$\mathcal{E}_i = \epsilon_{ijk}\langle u'_j b'_k\rangle = \epsilon_{ijk}\int \mathrm{d}^3k \int \mathrm{d}^3k'\langle \hat{u}_j(\mathbf{k};\,\tau)\hat{b}_k(\mathbf{k}';\,\tau)\rangle \mathrm{e}^{-\mathrm{i}(\mathbf{k}+\mathbf{k}')\cdot(\boldsymbol{\xi}-\langle\mathbf{U}\rangle\tau)}. \tag{5.34}$$

with the velocity–magnetic-field correlation tensor $\langle \hat{u}_j(\mathbf{k};\,\tau)\hat{b}_k(\mathbf{k}';\,\tau)\rangle$ calculated up to the first orders in the mean magnetic field $\langle\mathbf{B}\rangle$ and the scale-separation parameter δ,

$$\langle \hat{u}_j\hat{b}_k\rangle = \langle \hat{u}_{j\,00}\hat{b}_{k\,00}\rangle + \langle \hat{u}_{j\,01}\hat{b}_{k\,00}\rangle + \langle \hat{u}_{j\,00}\hat{b}_{k\,01}\rangle + \delta\langle \hat{u}_{j\,10}\hat{b}_{k\,00}\rangle + \delta\langle \hat{u}_{j\,00}\hat{b}_{k\,10}\rangle + \ldots, \tag{5.35}$$

where $\mathbf{u}_{01}$, $\mathbf{b}_{01}$, $\mathbf{u}_{10}$ and $\mathbf{b}_{10}$ are given in (5.23a–d). Utilizing (5.24) and (5.25a–d) we evaluate each of the first three contributions to $\langle \hat{u}_j\hat{b}_k\rangle$ separately,

$$\epsilon_{ijk}\int \mathrm{d}^3k \int \mathrm{d}^3k'\langle \hat{u}_{j\,00}(\mathbf{k};\,\tau)\hat{b}_{k\,00}(\mathbf{k}';\,\tau)\rangle \mathrm{e}^{-\mathrm{i}(\mathbf{k}+\mathbf{k}')\cdot(\boldsymbol{\xi}-\langle\mathbf{U}\rangle\tau)} = 0, \tag{5.36}$$

$$\begin{aligned}
&\epsilon_{ijk}\int \mathrm{d}^3k \int \mathrm{d}^3k'\langle \hat{u}_{j\,01}(\mathbf{k};\,\tau)\hat{b}_{k\,00}(\mathbf{k}';\,\tau)\rangle \mathrm{e}^{-\mathrm{i}(\mathbf{k}+\mathbf{k}')\cdot(\boldsymbol{\xi}-\langle\mathbf{U}\rangle\tau)}\\
&= -\mathrm{i}\langle B\rangle_m\epsilon_{ijk}\int_{-\infty}^{\tau}\mathrm{d}\tau'\int \mathrm{d}^3k P_{jq}(\mathbf{k})k_m G^{(uu)}(k;\,\tau,\,\tau')\times\\
&\qquad \int \mathrm{d}^3k'\langle \hat{b}_{q\,00}(\mathbf{k};\,\tau')\hat{b}_{k\,00}(\mathbf{k}';\,\tau)\rangle \mathrm{e}^{-\mathrm{i}(\mathbf{k}+\mathbf{k}')\cdot(\boldsymbol{\xi}-\langle\mathbf{U}\rangle\tau)}\\
&\quad -\mathrm{i}\langle B\rangle_m\epsilon_{ijk}\int_{-\infty}^{\tau}\mathrm{d}\tau'\int \mathrm{d}^3k P_{jq}(\mathbf{k})k_m G^{(ub)}(k;\,\tau,\,\tau')\times\\
&\qquad \int \mathrm{d}^3k'\langle \hat{u}_{q\,00}(\mathbf{k};\,\tau')\hat{b}_{k\,00}(\mathbf{k}';\,\tau)\rangle \mathrm{e}^{-\mathrm{i}(\mathbf{k}+\mathbf{k}')\cdot(\boldsymbol{\xi}-\langle\mathbf{U}\rangle\tau)}\\
&= \frac{1}{3}\langle B\rangle_i\int_{-\infty}^{\tau}\mathrm{d}\tau'\int \mathrm{d}^3k G^{(uu)}(k;\,\tau,\,\tau')H_{bb}(k;\,\tau,\,\tau')\\
&\quad +\frac{1}{3}\langle B\rangle_i\int_{-\infty}^{\tau}\mathrm{d}\tau'\int \mathrm{d}^3k G^{(ub)}(k;\,\tau,\,\tau')H_{bu}(k;\,\tau,\,\tau'),
\end{aligned} \tag{5.37}$$

$$\begin{aligned}
&\epsilon_{ijk}\int \mathrm{d}^3k \int \mathrm{d}^3k'\langle \hat{u}_{j\,00}(\mathbf{k};\,\tau)\hat{b}_{k\,01}(\mathbf{k}';\,\tau)\rangle \mathrm{e}^{-\mathrm{i}(\mathbf{k}+\mathbf{k}')\cdot(\boldsymbol{\xi}-\langle\mathbf{U}\rangle\tau)}\\
&=-\mathrm{i}\langle B\rangle_m\epsilon_{ijk}\int_{-\infty}^{\tau}\mathrm{d}\tau'\int \mathrm{d}^3k\times\\
&\quad \int \mathrm{d}^3k' P_{kq}(\mathbf{k}')k'_m G^{(bu)}(k';\,\tau,\,\tau')\langle \hat{u}_{j\,00}(\mathbf{k};\,\tau)\hat{b}_{q\,00}(\mathbf{k}';\,\tau')\rangle \mathrm{e}^{-\mathrm{i}(\mathbf{k}+\mathbf{k}')\cdot(\boldsymbol{\xi}-\langle\mathbf{U}\rangle\tau)}\\
&\quad -\mathrm{i}\langle B\rangle_m\epsilon_{ijk}\int_{-\infty}^{\tau}\mathrm{d}\tau'\int \mathrm{d}^3k\times\\
&\quad \int \mathrm{d}^3k' P_{kq}(\mathbf{k}')k'_m G^{(bb)}(k';\,\tau,\,\tau')\langle \hat{u}_{j\,00}(\mathbf{k};\,\tau)\hat{u}_{q\,00}(\mathbf{k}';\,\tau')\rangle \mathrm{e}^{-\mathrm{i}(\mathbf{k}+\mathbf{k}')\cdot(\boldsymbol{\xi}-\langle\mathbf{U}\rangle\tau)}\\
&=-\frac{1}{3}\langle B\rangle_i\int_{-\infty}^{\tau}\mathrm{d}\tau'\int \mathrm{d}^3k G^{(bu)}(k';\,\tau,\,\tau')H_{ub}(k;\,\tau,\,\tau')\\
&\quad -\frac{1}{3}\langle B\rangle_i\int_{-\infty}^{\tau}\mathrm{d}\tau'\int \mathrm{d}^3k G^{(bb)}(k';\,\tau,\,\tau')H_{uu}(k;\,\tau,\,\tau').
\end{aligned} \tag{5.38}$$

The last two contributions, namely $\delta\langle \hat{u}_{j\,10}\hat{b}_{k\,00}\rangle$ and $\delta\langle \hat{u}_{j\,00}\hat{b}_{k\,10}\rangle$ are evaluated as follows. First we use (5.23c and d) and (5.25a–d) to write

$$
\begin{aligned}
&\delta\epsilon_{ijk}\int d^3k\int d^3k'\langle \hat{u}_{j\,10}(\mathbf{k};\tau)\hat{b}_{k\,00}(\mathbf{k}';\tau)\rangle e^{-i(\mathbf{k}+\mathbf{k}')\cdot(\boldsymbol{\xi}-\langle \mathbf{U}\rangle\tau)}\\
&\quad= -\delta\epsilon_{ijk}\int_{-\infty}^{\tau}d\tau'\int d^3kP_{jn}(\mathbf{k})G^{(uu)}(k;\tau,\tau')\times\\
&\qquad\qquad\int d^3k'\left\langle \frac{\hat{D}\hat{u}_{n\,00}(\mathbf{k};\tau')}{DT_{\mathrm{I}}}\hat{b}_{k\,00}(\mathbf{k}';\tau)\right\rangle e^{-i(\mathbf{k}+\mathbf{k}')\cdot(\boldsymbol{\xi}-\langle \mathbf{U}\rangle\tau)}\\
&\quad-\delta\epsilon_{ijk}\int_{-\infty}^{\tau}d\tau'\int d^3kP_{jn}(\mathbf{k})G^{(ub)}(k;\tau,\tau')\times\\
&\qquad\qquad\int d^3k'\left\langle \frac{\hat{D}\hat{b}_{n\,00}(\mathbf{k};\tau')}{DT_{\mathrm{I}}}\hat{b}_{k\,00}(\mathbf{k}';\tau)\right\rangle e^{-i(\mathbf{k}+\mathbf{k}')\cdot(\boldsymbol{\xi}-\langle \mathbf{U}\rangle\tau)}\\
&\quad+\delta\frac{\partial\langle B\rangle_n}{\partial X_m}\epsilon_{ijk}\int_{-\infty}^{\tau}d\tau'\int d^3kP_{jn}(\mathbf{k})\times\\
&\qquad\int d^3k'[G^{(uu)}(k;\tau,\tau')\langle \hat{b}_{m\,00}(\mathbf{k};\tau')\hat{b}_{k\,00}(\mathbf{k}';\tau)\rangle\\
&\qquad\qquad-G^{(ub)}(k;\tau,\tau')\langle \hat{u}_{m\,00}(\mathbf{k};\tau')\hat{b}_{k\,00}(\mathbf{k}';\tau)\rangle]e^{-i(\mathbf{k}+\mathbf{k}')\cdot(\boldsymbol{\xi}-\langle \mathbf{U}\rangle\tau)}\\
&\quad-\delta\left(\frac{\partial\langle U\rangle_n}{\partial X_m}+\epsilon_{mn\ell}\Omega_{0\ell}\right)\epsilon_{ijk}\int_{-\infty}^{\tau}d\tau'\int d^3kP_{jn}(\mathbf{k})\times\\
&\qquad\int d^3k'[G^{(uu)}(k;\tau,\tau')\langle \hat{u}_{m\,00}(\mathbf{k};\tau')\hat{b}_{k\,00}(\mathbf{k}';\tau)\rangle\\
&\qquad\qquad-G^{(ub)}(k;\tau,\tau')\langle \hat{b}_{m\,00}(\mathbf{k};\tau')\hat{b}_{k\,00}(\mathbf{k}';\tau)\rangle]e^{-i(\mathbf{k}+\mathbf{k}')\cdot(\boldsymbol{\xi}-\langle \mathbf{U}\rangle\tau)}\\
&\quad+\delta\langle B\rangle_m\epsilon_{ijk}\int_{-\infty}^{\tau}d\tau'\int d^3kP_{jq}(\mathbf{k})\times\\
&\qquad\int d^3k'\left[G^{(ub)}(k;\tau,\tau')\left\langle \frac{\partial\hat{u}_{q\,00}(\mathbf{k};\tau')}{\partial X_{m\mathrm{I}}}\hat{b}_{k\,00}(\mathbf{k}';\tau)\right\rangle\right.\\
&\qquad\left.+G^{(uu)}(k;\tau,\tau')\left\langle \frac{\partial\hat{b}_{q\,00}(\mathbf{k};\tau')}{\partial X_{m\mathrm{I}}}\hat{b}_{k\,00}(\mathbf{k}';\tau)\right\rangle\right]e^{-i(\mathbf{k}+\mathbf{k}')\cdot(\boldsymbol{\xi}-\langle \mathbf{U}\rangle\tau)},
\end{aligned}
\tag{5.39}
$$

$$
\begin{aligned}
&\delta\epsilon_{ijk}\int d^3k\int d^3k'\langle\hat{u}_{j\,00}(\mathbf{k};\,\tau)\hat{b}_{k\,10}(\mathbf{k}';\,\tau)\rangle e^{-i(\mathbf{k}+\mathbf{k}')\cdot(\boldsymbol{\xi}-\langle\mathbf{U}\rangle\tau)}\\
&=-\delta\epsilon_{ijk}\int_{-\infty}^{\tau}d\tau'\int d^3k\times\\
&\quad\int d^3k'P_{kn}(\mathbf{k}')G^{(bu)}(k';\,\tau,\,\tau')\left\langle\hat{u}_{j\,00}(\mathbf{k};\,\tau)\frac{\hat{D}\hat{u}_{n\,00}(\mathbf{k}';\,\tau')}{DT_{\mathrm{I}}}\right\rangle e^{-i(\mathbf{k}+\mathbf{k}')\cdot(\boldsymbol{\xi}-\langle\mathbf{U}\rangle\tau)}\\
&-\delta\epsilon_{ijk}\int_{-\infty}^{\tau}d\tau'\int d^3k\times\\
&\quad\int d^3k'P_{kn}(\mathbf{k}')G^{(bb)}(k';\,\tau,\,\tau')\left\langle\hat{u}_{j\,00}(\mathbf{k};\,\tau)\frac{\hat{D}\hat{b}_{n\,00}(\mathbf{k}';\,\tau')}{DT_{\mathrm{I}}}\right\rangle e^{-i(\mathbf{k}+\mathbf{k}')\cdot(\boldsymbol{\xi}-\langle\mathbf{U}\rangle\tau)}\\
&+\delta\frac{\partial\langle B\rangle_n}{\partial X_m}\epsilon_{ijk}\int_{-\infty}^{\tau}d\tau'\int d^3k\times\\
&\quad\int d^3k'P_{kn}(\mathbf{k}')\Big[G^{(bu)}(k';\,\tau,\,\tau')\big\langle\hat{u}_{j\,00}(\mathbf{k};\,\tau)\hat{b}_{m\,00}(\mathbf{k}';\,\tau')\big\rangle\\
&\qquad-G^{(bb)}(k';\,\tau,\,\tau')\big\langle\hat{u}_{j\,00}(\mathbf{k};\,\tau)\hat{u}_{m\,00}(\mathbf{k}';\,\tau')\big\rangle\Big]e^{-i(\mathbf{k}+\mathbf{k}')\cdot(\boldsymbol{\xi}-\langle\mathbf{U}\rangle\tau)}\\
&-\delta\left(\frac{\partial\langle U\rangle_n}{\partial X_m}+\epsilon_{mn\ell}\Omega_{0\ell}\right)\epsilon_{ijk}\int_{-\infty}^{\tau}d\tau'\int d^3k\times\\
&\quad\int d^3k'P_{kn}(\mathbf{k}')\Big[G^{(bu)}(k';\,\tau,\,\tau')\big\langle\hat{u}_{j\,00}(\mathbf{k};\,\tau)\hat{u}_{m\,00}(\mathbf{k}';\,\tau')\big\rangle\\
&\qquad-G^{(bb)}(k';\,\tau,\,\tau')\big\langle\hat{u}_{j\,00}(\mathbf{k};\,\tau)\hat{b}_{m\,00}(\mathbf{k}';\,\tau')\big\rangle\Big]e^{-i(\mathbf{k}+\mathbf{k}')\cdot(\boldsymbol{\xi}-\langle\mathbf{U}\rangle\tau)}\\
&+\delta\langle B\rangle_m\epsilon_{ijk}\int_{-\infty}^{\tau}d\tau'\int d^3k\times\\
&\quad\int d^3k'P_{kq}(\mathbf{k}')\left[G^{(bb)}(k';\,\tau,\,\tau')\left\langle\hat{u}_{j\,00}(\mathbf{k};\,\tau)\frac{\partial\hat{u}_{q\,00}(\mathbf{k}';\,\tau')}{\partial X_{m\mathrm{I}}}\right\rangle\right.\\
&\qquad\left.+G^{(bu)}(k';\,\tau,\,\tau')\left\langle\hat{u}_{j\,00}(\mathbf{k};\,\tau)\frac{\partial\hat{b}_{q\,00}(\mathbf{k}';\,\tau')}{\partial X_{m\mathrm{I}}}\right\rangle\right]e^{-i(\mathbf{k}+\mathbf{k}')\cdot(\boldsymbol{\xi}-\langle\mathbf{U}\rangle\tau)}.
\end{aligned}
\tag{5.40}
$$

Next, with the aid of (5.24), (5.12) and (5.8) we arrive at

$$
\begin{aligned}
&\delta\epsilon_{ijk}\int \mathrm{d}^3k \int \mathrm{d}^3k' \langle \hat{u}_{j\,10}(\mathbf{k};\,\tau)\hat{b}_{k\,00}(\mathbf{k}';\,\tau)\rangle \mathrm{e}^{-\mathrm{i}(\mathbf{k}+\mathbf{k}')\cdot(\boldsymbol{\xi}-\langle\mathbf{U}\rangle\tau)} \\
&= \frac{2}{3}\Omega_i \int_{-\infty}^{\tau} \mathrm{d}\tau' \int \mathrm{d}^3k \left[G^{(uu)}(k;\,\tau,\,\tau')Q_{bu}(k;\,\tau,\,\tau') + G^{(ub)}(k;\,\tau,\,\tau')Q_{bb}(k;\,\tau,\,\tau') \right] \\
&\quad - \frac{1}{3}\epsilon_{ijk}\frac{\partial\langle B\rangle_k}{\partial x_j} \int_{-\infty}^{\tau} \mathrm{d}\tau' \int \mathrm{d}^3k \Big[G^{(uu)}(k;\,\tau,\,\tau')Q_{bb}(k;\,\tau,\,\tau') \\
&\qquad\qquad - G^{(ub)}(k;\,\tau,\,\tau')Q_{bu}(k;\,\tau,\,\tau') \Big] \\
&\quad + \frac{1}{3}\left(\epsilon_{ijk}\frac{\partial\langle U\rangle_k}{\partial x_j} + 2\Omega_i \right) \int_{-\infty}^{\tau} \mathrm{d}\tau' \int \mathrm{d}^3k \Big[G^{(uu)}(k;\,\tau,\,\tau')Q_{bu}(k;\,\tau,\,\tau') \\
&\qquad\qquad - G^{(ub)}(k;\,\tau,\,\tau')Q_{bb}(k;\,\tau,\,\tau') \Big] + \ldots,
\end{aligned}
\tag{5.41}
$$

$$
\begin{aligned}
&\epsilon_{ijk}\int \mathrm{d}^3k \int \mathrm{d}^3k' \langle \hat{u}_{j\,00}(\mathbf{k};\,\tau)\hat{b}_{k\,10}(\mathbf{k}';\,\tau)\rangle \mathrm{e}^{-\mathrm{i}(\mathbf{k}+\mathbf{k}')\cdot(\boldsymbol{\xi}-\langle\mathbf{U}\rangle\tau)} \\
&= -\frac{2}{3}\Omega_i \int_{-\infty}^{\tau} \mathrm{d}\tau' \int \mathrm{d}^3k \left[G^{(bu)}(k;\,\tau,\,\tau')Q_{uu}(k;\,\tau,\,\tau') + G^{(bb)}(k;\,\tau,\,\tau')Q_{ub}(k;\,\tau,\,\tau') \right] \\
&\quad - \frac{1}{3}\epsilon_{ijk}\frac{\partial\langle B\rangle_k}{\partial x_j} \int_{-\infty}^{\tau} \mathrm{d}\tau' \int \mathrm{d}^3k \Big[G^{(bb)}(k;\,\tau,\,\tau')Q_{uu}(k;\,\tau,\,\tau') \\
&\qquad\qquad - G^{(bu)}(k;\,\tau,\,\tau')Q_{ub}(k;\,\tau,\,\tau') \Big] \\
&\quad + \frac{1}{3}\left(\epsilon_{ijk}\frac{\partial\langle U\rangle_k}{\partial x_j} + 2\Omega_i \right) \int_{-\infty}^{\tau} \mathrm{d}\tau' \int \mathrm{d}^3k \Big[G^{(bb)}(k;\,\tau,\,\tau')Q_{ub}(k;\,\tau,\,\tau') \\
&\qquad\qquad - G^{(bu)}(k;\,\tau,\,\tau')Q_{uu}(k;\,\tau,\,\tau') \Big] + \ldots,
\end{aligned}
\tag{5.42}
$$

where the dots indicate the presence of some additional, less standard terms (for more details, the interested reader is referred to [1, 3]). Finally, by (5.35), (5.36–5.38) and (5.41–5.42) the large-scale EMF can be cast in the following form

$$
\langle \mathbf{u}' \times \mathbf{b}' \rangle = \alpha\langle \mathbf{B} \rangle - \beta\nabla \times \langle \mathbf{B} \rangle + \gamma\langle \mathbf{W} \rangle + 2\tilde{\gamma}\boldsymbol{\Omega} + \ldots, \tag{5.43}
$$

where the transport coefficients α, β and γ, $\tilde{\gamma}$ are given by

$$
\alpha = \frac{1}{3}[-I\{G_{bb},\, H_{uu}\} + I\{G_{uu},\, H_{bb}\} - I\{G_{bu},\, H_{ub}\} + I\{G_{ub},\, H_{bu}\}], \tag{5.44}
$$

$$
\beta = \frac{1}{3}\left[I\{G_{bb},\, Q_{uu}\} + I\{G_{uu},\, Q_{bb}\} - I\{G_{bu},\, Q_{ub}\} - I\{G_{ub},\, Q_{bu}\} \right], \tag{5.45}
$$

$$
\gamma = \frac{1}{3}\left[I\{G_{bb},\, Q_{ub}\} + I\{G_{uu},\, Q_{bu}\} - I\{G_{bu},\, Q_{uu}\} - I\{G_{ub},\, Q_{bb}\} \right], \tag{5.46}
$$

$$\tilde{\gamma} = \frac{2}{3}\left[I\{G_{uu}, Q_{bu}\} - I\{G_{bu}, Q_{uu}\}\right], \tag{5.47}$$

with

$$I\{A, B\} = \int d^3k \int_{-\infty}^{\tau} d\tau' A(k; \tau, \tau')B(k; \tau, \tau'), \tag{5.48}$$

dependent on the evolution history. Recall that $\nabla \times \langle \mathbf{B} \rangle = \langle \mathbf{J} \rangle$ and $\langle \mathbf{W} \rangle = \nabla \times \langle \mathbf{U} \rangle$ denote the large-scale electric currents and the large-scale vorticity, respectively.

We now concentrate on the α-effect, which can be decomposed into two contributions,

$$\alpha = \alpha_R + \alpha_X, \tag{5.49}$$

that is one related to the residual helicity $\langle \mathbf{u}' \cdot \mathbf{w}' - \mathbf{b}' \cdot \mathbf{j}' \rangle$, i.e.

$$\begin{aligned}\alpha_R = \frac{1}{3} \int d^3k \int_{-\infty}^{\tau} d\tau' [&G_{uu}(k, \mathbf{X}; \tau, \tau', T)H_{bb}(k, \mathbf{X}; \tau, \tau', T) \\ &- G_{bb}(k, \mathbf{X}; \tau, \tau', T)H_{uu}(k, \mathbf{X}; \tau, \tau', T)],\end{aligned} \tag{5.50}$$

and a less obvious one, related to the cross-helicity, $\langle \mathbf{u}' \cdot \mathbf{b}' \rangle$, and the quantity $\langle \mathbf{u}' \cdot \mathbf{j}' \rangle$ which takes the form

$$\begin{aligned}\alpha_X = -\frac{1}{3} \int d^3k \int_{-\infty}^{\tau} d\tau' [&G_{bu}(k, \mathbf{X}; \tau, \tau', T)H_{ub}(k, \mathbf{X}; \tau, \tau', T) \\ &- G_{ub}(k, \mathbf{X}; \tau, \tau', T)H_{bu}(k, \mathbf{X}; \tau, \tau', T)].\end{aligned} \tag{5.51}$$

Since the helical functions of the background turbulence satisfy

$$H_{bu}(k; \tau, \tau') = H_{ub}(k; \tau', \tau), \tag{5.52}$$

we obtain

$$\begin{aligned}\alpha_X = -\frac{1}{3} \int d^3k \int_{-\infty}^{\tau} d\tau' [&G_{bu}(k; \tau, \tau')H_{ub}(k; \tau, \tau') \\ &- G_{ub}(k; \tau, \tau')H_{ub}(k; \tau', \tau)],\end{aligned} \tag{5.53}$$

where, as before, the explicit dependence on the slow variables $\mathbf{X}$ and T was suppressed in notation for clarity. We now introduce the following symmetric and antisymmetric parts of H_{ub} with respect to exchange of the time variables

$$H_{ub}^{(s)}(k; \tau, \tau') = \frac{1}{2}(H_{ub}(k; \tau, \tau') + H_{ub}(k; \tau', \tau)), \tag{5.54}$$

$$H_{ub}^{(a)}(k; \tau, \tau') = \frac{1}{2}(H_{ub}(k; \tau, \tau') - H_{ub}(k; \tau', \tau)), \tag{5.55}$$

which allows one to further separate the α_X term into two contributions

$$\alpha_X = -\frac{1}{3}\int d^3k \int_{-\infty}^{\tau} d\tau'[G_{ub}(k;\tau,\tau') + G_{bu}(k;\tau,\tau')]H_{ub}^{(a)}(k;\tau,\tau') \\ +\frac{1}{3}\int d^3k \int_{-\infty}^{\tau} d\tau'[G_{ub}(k;\tau,\tau') - G_{bu}(k;\tau,\tau')]H_{ub}^{(s)}(k;\tau,\tau'). \tag{5.56}$$

The first term in equation (5.56), i.e.

$$\alpha_{X\,\mathrm{neq}} \stackrel{\mathrm{def}}{=} -\frac{1}{3}\int d^3k \int_{-\infty}^{\tau} d\tau'[G_{ub}(k;\tau,\tau') + G_{bu}(k;\tau,\tau')]H_{ub}^{(a)}(k;\tau,\tau'), \tag{5.57}$$

clearly constitutes a contribution from non-stationarity of the turbulence, as the antisymmetric part $H_{ub}^{(a)}$ is clearly a non-equilibrium effect.

5.2.1 Physical features of the non-equilibrium $\alpha_{X\,\mathrm{neq}}$-effect

If we further assume that the function

$$\mathcal{G}(\tau,\tau') \stackrel{\mathrm{def}}{=} G_{ub}(\tau,\tau') + G_{bu}(\tau,\tau'), \tag{5.58}$$

is independent of k, the non-equilibrium α-effect can be expressed as follows,

$$\alpha_{X\,\mathrm{neq}} = -\frac{1}{3}\int_{-\infty}^{\tau} d\tau'\mathcal{G}(\tau,\tau')\langle \mathbf{u}'_{00}\cdot \mathbf{j}'_{00}\rangle^{(a)}(\mathbf{x},\tau,\tau'), \tag{5.59}$$

where

$$\langle \mathbf{u}'_{00}\cdot \mathbf{j}'_{00}\rangle^{(a)}(\mathbf{x},\tau,\tau') = \frac{1}{2}\left[\langle \mathbf{u}'_{00}(\mathbf{x},\tau)\cdot \mathbf{j}'_{00}(\mathbf{x},\tau')\rangle - \langle \mathbf{u}'_{00}(\mathbf{x},\tau')\cdot \mathbf{j}'_{00}(\mathbf{x},\tau)\rangle\right]. \tag{5.60}$$

The memory effect, expressed by the time integral in (5.60) is clearly crucial, as $\langle \mathbf{u}'_{00}\cdot \mathbf{j}'_{00}\rangle^{(a)}(\mathbf{x},\tau,\tau) = 0$.

The second term in (5.56) is likely to be small because of the factor $G_{ub}(k;\tau,\tau') - G_{bu}(k;\tau,\tau')$. For example, as mentioned at the end of section 5.1.2, in the case when $\nu = \eta$ the two response functions G_{ub} and G_{bu} are equal and $\alpha_X = \alpha_{X\,\mathrm{neq}}$. This still holds approximately true, when the diffusivities are unequal but weak,

$$G_{ub} \approx G_{bu}, \qquad \text{and} \qquad \alpha_X \approx \alpha_{X\,\mathrm{neq}}. \tag{5.61}$$

Next, inspection of the evolution equations for the Green's functions leads to the conclusion that G_{ub} must be an odd function of $\mathbf{b}'_{00}$. This is expected, since the α_X contribution to the α-effect results from action of the Lorentz force, and since H_{ub} is associated with the quantity $\left\langle \mathbf{u}'_{00}\cdot \mathbf{j}'_{00}\right\rangle$, i.e. H_{ub} is linear in $\mathbf{b}'_{00}$, it follows that G_{ub} must be an odd function of the latter. Moreover, since $\left\langle \mathbf{u}'_{00}\cdot \mathbf{j}'_{00}\right\rangle$ is a scalar quantity (does not change sign under reflections), G_{ub} must be skew. The only dynamical quantity that is skew and odd in $\mathbf{b}'_{00}$ is the cross-helicity, $\langle \mathbf{u}'_{00}\cdot \mathbf{b}'_{00}\rangle$, hence we expect that $G_{ub} \propto Q_{ub}$. Having in mind that the response function $\mathcal{G}(\tau,\tau')$ is non-dimensional we can now provide the following estimate of the non-equilibrium $\alpha_{X\,\mathrm{neq}}$-effect

$$\alpha_{\mathrm{X\,neq}} \approx -\frac{2}{3}\int_{-\infty}^{\tau} \mathrm{d}\tau' \Upsilon^{(s)}(\mathbf{x}, \tau, \tau')\langle \mathbf{u}'_{00} \cdot \mathbf{j}'_{00}\rangle^{(a)}(\mathbf{x}, \tau, \tau'), \tag{5.62}$$

where

$$\Upsilon(\mathbf{x}, \tau, \tau') = \frac{\langle \mathbf{u}'_{00}(\mathbf{x}, \tau) \cdot \mathbf{b}'_{00}(\mathbf{x}, \tau')\rangle}{\sqrt{\langle u'^2_{00}\rangle(\mathbf{x}, \tau)\langle b'^2_{00}\rangle(\mathbf{x}, \tau')}}, \tag{5.63}$$

$$\Upsilon^{(s)}(\mathbf{x}, \tau, \tau') = \frac{1}{2}[\Upsilon(\mathbf{x}, \tau, \tau') + \Upsilon(\mathbf{x}, \tau', \tau)], \tag{5.64}$$

and the cross-helicity has been normalized by the geometric mean of the kinetic and magnetic fluctuational energies (see also [11] for a discussion of different cross-helicity normalizations). The latter equation expresses an effect which results from the lack of equilibrium in the turbulent state. A very rough estimate of the $\alpha_{\mathrm{X\,neq}}$ coefficient could be obtained by replacing $\Upsilon^{(s)}(\mathbf{x}, \tau, \tau')$ with $\Upsilon(\mathbf{x}, \tau, \tau)$, which would lead to

$$\begin{aligned}\alpha_{\mathrm{X\,neq}} \approx -\frac{1}{3}\frac{\langle \mathbf{u}'_{00} \cdot \mathbf{b}'_{00}\rangle}{\sqrt{\langle u'^2_{00}\rangle\langle b'^2_{00}\rangle}}\int_{-\infty}^{\tau} \mathrm{d}\tau' \Big[&\langle \mathbf{u}'_{00}(\mathbf{x}, \tau) \cdot \mathbf{j}'_{00}(\mathbf{x}, \tau')\rangle \\ &-\langle \mathbf{u}'_{00}(\mathbf{x}, \tau') \cdot \mathbf{j}'_{00}(\mathbf{x}, \tau)\rangle\Big].\end{aligned} \tag{5.65}$$

The latter equation, although approximate only, has the advantage of simplicity and bears some similarity to the commonly used in the literature estimate of the stationary α_{R}-effect, namely that $\alpha_{\mathrm{R}} \approx -\tau_t(\langle \mathbf{u}' \cdot \mathbf{w}'\rangle - \langle \mathbf{b}' \cdot \mathbf{j}'\rangle)/3$, where τ_t denotes the turn over time of the most energetic turbulent eddies.

Finally, we also expect the $\langle \mathbf{u}'_{00} \cdot \mathbf{j}'_{00}\rangle$ correlations in fully turbulent flows to be proportional to the kinetic helicity $\langle \mathbf{u}'_{00} \cdot \mathbf{w}'_{00}\rangle$, since typically the velocities and magnetic fields tend to align in such flows. Again, the prefactor must be skew and odd in $\mathbf{b}'_{00}$, therefore the following proposition seems reasonable

$$\langle \mathbf{u}'_{00} \cdot \mathbf{j}'_{00}\rangle \approx \Upsilon\langle \mathbf{u}'_{00} \cdot \mathbf{w}'_{00}\rangle = \frac{\langle \mathbf{u}'_{00} \cdot \mathbf{b}'_{00}\rangle\langle \mathbf{u}'_{00} \cdot \mathbf{w}'_{00}\rangle}{\sqrt{\langle u'^2_{00}\rangle\langle b'^2_{00}\rangle}}. \tag{5.66}$$

Introducing the latter relation into (5.62) leads to

$$\alpha_{\mathrm{X\,neq}} \approx -\frac{2}{3}\int_{-\infty}^{\tau} \mathrm{d}\tau' (\Upsilon^{(s)}(\mathbf{x}, \tau, \tau'))^2\langle \mathbf{u}'_{00} \cdot \mathbf{w}'_{00}\rangle^{(a)}(\mathbf{x}, \tau, \tau'), \tag{5.67}$$

which shows, that the non-equilibrium $\alpha_{\mathrm{X\,neq}}$-effect relies on coexistence of the kinetic and cross-helicities and their history in the MHD turbulence (more precisely, in the case of kinetic helicity only the antisymmetric part of the time correlations $\langle \mathbf{u}'_{00} \cdot \mathbf{w}'_{00}\rangle^{(a)}(\mathbf{x}, \tau, \tau')$ contributes to the non-equilibrium effect). In fact, the proportionality between $\langle \mathbf{u}'_{00} \cdot \mathbf{j}'_{00}\rangle$ and the cross-helicity $\langle \mathbf{u}'_{00} \cdot \mathbf{b}'_{00}\rangle$, in line with the presence of the coefficient Υ in (5.66), can be easily shown within the first order smoothing approximation which was utilized in chapter 3, when the turbulent fluctuations of velocity and magnetic fields are governed by

$$(-\mathrm{i}\omega + \nu k^2)\hat{u}_i(\mathbf{q}) + 2\Omega\epsilon_{i3j}\hat{u}_j(\mathbf{q}) = \hat{f}_i(\mathbf{q}) + \mathrm{i}\mathbf{k}\cdot\langle\mathbf{B}\rangle\hat{b}_i(\mathbf{q}), \tag{5.68a}$$

$$(-\mathrm{i}\omega + \eta k^2)\hat{b}_i(\mathbf{q}) = \mathrm{i}\mathbf{k}\cdot\langle\mathbf{B}\rangle\hat{u}_i(\mathbf{q}), \tag{5.68b}$$

and we can adopt the same stochastic forcing $\mathbf{f}$ as earlier, defined by the correlation function

$$\left\langle\hat{f}_i(\mathbf{k},\omega)\hat{f}_j(\mathbf{k}',\omega')\right\rangle = \left[\frac{D_0}{k^3}P_{ij}(\mathbf{k}) + \mathrm{i}\frac{D_1}{k^5}\epsilon_{ijk}k_k\right]\times \\ \delta(\mathbf{k}+\mathbf{k}')\frac{\delta(\omega+\omega'+\tilde{\omega}) - \delta(\omega+\omega'-\tilde{\omega})}{2\mathrm{i}}. \tag{5.69}$$

It follows that in the limit of weak seed field, $\langle\mathbf{B}\rangle^2 \ll \langle\mathbf{U}\rangle^2$, we have $\hat{u}_i(\mathbf{q}) \approx \mathfrak{G}_{ij}\hat{f}_j(\mathbf{q})$ and $\hat{b}_i(\mathbf{q}) \approx \mathrm{i}\mathbf{k}\cdot\langle\mathbf{B}\rangle\mathfrak{G}_{ij}\hat{f}_j(\mathbf{q})/(-\mathrm{i}\omega + \eta k^2)$, where the operator $\mathfrak{G}$ is defined as

$$\mathfrak{G}_{ij} = \frac{1}{(-\mathrm{i}\omega + \nu k^2)^2 + 4\Omega^2\frac{k_z^2}{k^2}}\left[(-\mathrm{i}\omega + \nu k^2)\delta_{ij} - 2\Omega\epsilon_{i3j} + 2\Omega\frac{k_ik_m}{k^2}\epsilon_{jm3}\right], \tag{5.70}$$

and the correlation functions $\langle u_i'(\mathbf{x},t)b_i'(\mathbf{x},t)\rangle$ and $\left\langle u_i'(\mathbf{x},t)j_i'(\mathbf{x},t)\right\rangle$ can be calculated in a straightforward manner (with the aid of (3.99a and b)), which leads to

$$\left\langle u_i'(\mathbf{x},t)j_i'(\mathbf{x},t)\right\rangle \approx -\frac{\pi^2\tilde{\omega}D_0}{\Omega^2\nu^2k_\ell^2}(\langle\mathbf{B}\rangle\cdot\boldsymbol{\Omega}) \approx -\frac{2D_0k_\ell^2}{D_1}\langle u_i'(\mathbf{x},t)b_i'(\mathbf{x},t)\rangle, \tag{5.71}$$

and a similar result is obtained for stationary turbulence. This gives additional justification for the choice of the proportionality coefficient Υ in (5.66) between $\langle\mathbf{u}_{00}'\cdot\mathbf{j}_{00}'\rangle$ and the kinetic helicity.

In principle, the same symmetry arguments as in the case of $\alpha_{\mathrm{X\,neq}}$ can be also applied to the second term in (5.56) which is therefore proportional to the non-dimensional cross-helicity $\Upsilon = \Upsilon(\mathbf{x},\tau,\tau)$ and the quantity $\langle\mathbf{u}_{00}'(\mathbf{x},\tau)\cdot\mathbf{j}_{00}'(\mathbf{x},\tau)\rangle$, i.e. $\alpha_{\mathrm{X}} - \alpha_{\mathrm{X\,neq}} \propto \Upsilon^2\langle\mathbf{u}_{00}'\cdot\mathbf{w}_{00}'\rangle$. However, as remarked above this effect should be weak, when the diffusion is weak or the magnetic Prandtl number $Pm = \nu/\eta \approx 1$.

5.2.2 Further calculation of the $\alpha_{\mathrm{X\,neq}}$-effect; Taylor (1921) ansatz for correlation functions

We will now investigate this dynamo mechanism in some more detail. In order to calculate the effect of non-equilibrium turbulence we adopt a similar approach to that in section 6 of [3], based on a standard type of ansatz used in turbulence modelling introduced by Taylor (1921) [12] and Tennekes (1979) [13]. Here this model is used to show explicitly how non-stationarity generates the $\alpha_{\mathrm{X\,neq}}$-correction and emphasise the role of the dynamical history of the helicities. In stationary turbulence the functions $H_{fg}(k, \mathbf{X}; \tau, \tau_1, T)$ and $G_{fg}(k, \mathbf{X}; \tau, \tau_1, T)$ depend only on $|\tau - \tau_1|$, hence in order to study the non-equilibrium effects we postulate formulae for these functions similar to those of Yoshizawa (1998) [3] (cf formulae (6.53) and (6.54) in that book), but modified in order to introduce simple explicit and distinct dependencies on τ and τ_1 (cf also [7])

$$H_{fg}(k, \mathbf{k}\cdot\langle\mathbf{B}\rangle, \mathbf{X}; \tau, \tau_1, T) = \sigma(k, \mathbf{X}, T)\mathrm{e}^{-\varpi(k, \mathbf{X}, T)|\tau-\tau_1|}\mathcal{H}(\tau)\mathcal{H}_1(\tau_1), \tag{5.72a}$$

$$G_{rs}(k, \mathbf{X}; \tau, \tau_1, T) = \theta(\tau - \tau_1)\varsigma(k, \mathbf{X}, T)\mathrm{e}^{-\varpi(k, \mathbf{X}, T)|\tau-\tau_1|}\mathcal{G}(\tau)\mathcal{G}_1(\tau_1), \tag{5.72b}$$

for some functions $\mathcal{H}(\tau)$, $\mathcal{H}_1(\tau_1)$, $\mathcal{G}(\tau)$ and $\mathcal{G}_1(\tau_1)$. The following calculation is correct for all types of correlations, i.e. for f, g and r, s representing either $\mathbf{u}$ or $\mathbf{b}$, however, since we are interested in the $\alpha_{\mathrm{X\,neq}}$-effect we will be primarily focused on the case $f = u$, $g = b$ and $r = b$, $s = u$.

We can decompose the functions $\mathcal{H}(\tau)$, $\mathcal{H}_1(\tau_1)$, $\mathcal{G}(\tau)$ and $\mathcal{G}_1(\tau_1)$ into Fourier modes, which allows us to adopt the following, simpler, generic model

$$H_{fg}(\tau, \tau_1) = \sigma\mathrm{e}^{-\varpi|\tau-\tau_1|}\sin(\varpi_{h0}\tau)\sin(\varpi_{h1}\tau_1), \tag{5.73}$$

$$G_{rs}(\tau, \tau_1) = \theta(\tau - \tau_1)\varsigma\mathrm{e}^{-\varpi|\tau-\tau_1|}\sin(\varpi_{g0}\tau)\sin(\varpi_{g1}\tau_1), \tag{5.74}$$

where the dependence on the slow variables and also additionally on the wave number k was suppressed in notation for clarity ($\theta(\tau - \tau_1)$ is the Heaviside unit step function); moreover $\varpi > 0$ and to fix ideas we also assume $\varpi_{h0} > 0$, $\varpi_{g0} > 0$, $\varpi_{h1} > 0$ and $\varpi_{g1} > 0$. For the sake of simplicity we consider the case

$$G_{ub} \approx G_{bu}, \tag{5.75}$$

which, as explained above, corresponds to the case of very weak diffusion effects or $Pm = 1$. The following calculation

$$\begin{aligned}\int_{-\infty}^{\tau} & \mathrm{d}\tau_1 G_{rs}(\tau, \tau_1)H_{fg}(\tau, \tau_1)\\ &= \frac{\sigma\varsigma}{4}(\cos\Delta_0\tau - \cos\Sigma_0\tau)\left[\frac{1}{4\varpi^2 + \Delta_1^2}(2\varpi\cos\Delta_1\tau + \Delta_1\sin\Delta_1\tau)\right.\\ &\qquad \left. - \frac{1}{4\varpi^2 + \Sigma_1^2}(2\varpi\cos\Sigma_1\tau + \Sigma_1\sin\Sigma_1\tau)\right],\end{aligned} \tag{5.76}$$

where

$$\Delta_i = \varpi_{hi} - \varpi_{gi}, \qquad \Sigma_i = \varpi_{hi} + \varpi_{gi}, \tag{5.77}$$

shows, that in non-equilibrium turbulence both contributions to the α-effect, the 'standard' α_{R} (obtained for $f = g = u$, $r = s = b$ and for $f = g = b$, $r = s = u$) and the one associated with cross-helicity α_{X} (obtained for $f = s = u$, $g = r = b$ and for $f = s = b$, $g = r = u$), are enhanced by non-stationarity. Since the frequencies correspond to the fast oscillations of turbulent fluctuations, in most of the cases the cosines and sines do not contribute to large time-scales (their time average vanishes). Under the time average over long time-scales $\delta^{-1}t$ the non-zero contribution comes from the cases $\varpi_{hi} = \varpi_{gi}$ (or $\varpi_{hi} \approx \varpi_{gi}$). Therefore, we pick $(\varpi, \varpi_h, \varpi_g)$-modes such that the following relations are satisfied

$$\Delta_i \ll \varpi \ll \varpi_{hi}, \varpi_{gi}, \quad \text{for} \quad i = 0, 1, \tag{5.78}$$

in which case

$$\int_{-\infty}^{\tau} d\tau_1 G_{ub}(\tau, \tau_1) H_{ub}(\tau, \tau_1) \approx \frac{\sigma\varsigma}{8\varpi}, \tag{5.79}$$

where now $\sigma(k, X, T)$ is the spectral function corresponding to the $\langle \mathbf{u}' \cdot \mathbf{j}' \rangle$ correlations, and $\varsigma(k, X, T)$ defines the cross-interactions response function G_{ub}; for comparison in the stationary case one obtains $\sigma_s\varsigma_s/2\varpi_s$ with $H_{fg} = \sigma_s \exp(-\varpi_s \mid \tau - \tau_1 \mid)$, $G_{fg} = \varsigma_s \exp(-\varpi_s \mid \tau - \tau_1 \mid)$.

We note that the influence of non-stationarity on the 'standard' $\alpha_{\rm R}$ contribution, i.e. using this section's terminology the $\alpha_{\rm R\,neq}$-effect (obtained for $f = g = u$, $r = s = b$ and for $f = g = b$, $r = s = u$), has been studied in previous chapters and the relation (5.78) clearly corresponds to the effect of beating waves inducing non-stationarity in the turbulent wave field, discussed above. Here we concentrate on the cross-helicity contribution $\alpha_{\rm X} \approx \alpha_{\rm X\,neq}$, which is apparent within the TSDIA approach. Introduction of the formulae (5.73 and 5.74) into (5.57) yields

$$\alpha_{\rm X\,neq} \approx -\frac{\pi}{6} \int dk \frac{\sigma\varsigma k^2}{\varpi}. \tag{5.80}$$

According to our previous observations in the above, we have $\varsigma \sim \Upsilon$. It is also of interest to note that a very similar result is obtained if one assumes a simpler non-stationary form of the H_{ub} and G_{ub} functions

$$H_{ub}(\tau, \tau_1) = \sigma e^{-\varpi|\tau - \tau_1|} \sin\left[\varpi_h(\tau - \tau_1)\right], \tag{5.81}$$

$$G_{ub}(\tau, \tau_1) = \theta(\tau - \tau_1)\varsigma e^{-\varpi|\tau - \tau_1|} \sin\left[\varpi_g(\tau - \tau_1)\right], \tag{5.82}$$

which satisfies time antisymmetry condition $H_{ub}(\tau, \tau_1) = -H_{ub}(\tau_1, \tau)$, and considers the limit (5.78).

In the above calculation we have used some standard models of the statistical properties of turbulence in order to emphasize the importance of the history of evolution of the helicities in the turbulent dynamo process. The $\alpha_{\rm X\,neq}$-effect, induced by the simultaneous presence of cross- and kinetic helicities, can be strong and depends on their magnitude. All the three ingredients are crucial: (i) cross-helicity, (ii) kinetic helicity and (iii) non-stationarity. Clearly, in turbulence which possesses the cross- and kinetic helicities in abundance and which is out of equilibrium, a correction to the standard α-effect is created, which involves the cross-helicity and memory effects.

5.3 Coexistence of the kinetic and cross-helicities in turbulence

We now consider the question of the likelihood of coexistence of the cross and kinetic helicities in developed turbulence. Although it is not possible to draw definite conclusions in this matter, it is still instructive to study the sources and sinks of the cross-helicity in turbulent flows in order to develop some intuition about its generation.

In (5.71) we showed that under the first-order smoothing approximation the cross-helicity is proportional to the product $\langle \mathbf{B} \rangle \cdot \mathbf{\Omega}$. In other words, within the First-Order Smoothing Approximation (FOSA) approach the existence of the cross-helicity is dependent on the existence of the mean field component parallel to the background rotation vector. A more general approach can be based on the cross-helicity evolution equation (2.55), also provided, e.g. in [4, 11, 14–16]. This equation involves mean quantities such as the mean EMF $\boldsymbol{\mathcal{E}}$ and the mean Reynolds and Maxwell stresses $\langle u_i' u_j' - b_i' b_j' \rangle$. For the former we utilize the result (5.43) and for the latter we take the expression obtained also via the TSDIA approach in [16], i.e.

$$-\langle u_i' u_j' - b_i' b_j' \rangle \frac{\partial \langle B \rangle_i}{\partial x_j} = \frac{7}{10} \beta \mathcal{S}_{ij} \mathcal{M}_{ij} - \frac{7}{10} \gamma \,\mathrm{Tr}\,(\mathcal{M}^2), \tag{5.83}$$

where

$$\mathcal{S}_{ij} = \frac{\partial \langle U \rangle_i}{\partial x_j} + \frac{\partial \langle U \rangle_j}{\partial x_i}, \qquad \mathcal{M}_{ij} = \frac{\partial \langle B \rangle_i}{\partial x_j} + \frac{\partial \langle B \rangle_j}{\partial x_i}, \tag{5.84}$$

are the symmetric parts of the mean velocity and magnetic field gradient tensors. This leads to

$$\begin{aligned} \frac{D}{Dt} \langle \mathbf{u}' \cdot \mathbf{b}' \rangle = &- \alpha(\langle \mathbf{B} \rangle \cdot \langle \mathbf{W} \rangle + 2 \langle \mathbf{B} \rangle \cdot \mathbf{\Omega}) + \beta(\langle \mathbf{J} \rangle \cdot \langle \mathbf{W} \rangle + 2 \langle \mathbf{J} \rangle \cdot \mathbf{\Omega}) \\ &- \gamma(\langle \mathbf{W} \rangle + 2\mathbf{\Omega})^2 + 2(\gamma - \tilde{\gamma})(\langle \mathbf{W} \rangle \cdot \mathbf{\Omega} + 2\Omega^2) \\ &- \frac{7}{10} \gamma \,\mathrm{Tr}\,(\mathcal{M}^2) + \frac{7}{10} \beta \,\mathrm{Tr}\,(\mathcal{S} \cdot \mathcal{M}) + (\nabla \zeta \times \langle \mathbf{B} \rangle) \cdot (\langle \mathbf{W} \rangle + 2\mathbf{\Omega}) \\ &+ \nabla \cdot \left[\left\langle \left(-\Pi' + \frac{\mathbf{u}'^2 + \mathbf{b}'^2}{2} \right) \mathbf{b}' \right\rangle + \left\langle \frac{\mathbf{u}'^2 + \mathbf{b}'^2}{2} \right\rangle \langle \mathbf{B} \rangle \right. \\ &\qquad \left. - \nu \langle \mathbf{w}' \times \mathbf{b}' \rangle + \eta \langle \mathbf{u}' \times \mathbf{j}' \rangle \right] - (\nu + \eta) \langle \mathbf{w}' \cdot \mathbf{j}' \rangle. \end{aligned} \tag{5.85}$$

Of course if the turbulence is stirred with some forcing $\mathbf{f}$ there is yet another production term $\langle \mathbf{f} \cdot \mathbf{b}' \rangle$.

According to (5.67) the magnitude of the non-equilibrium α-effect depends on both the kinetic and cross-helicities and their history. The total α-effect consists of the two contributions $\alpha = \alpha_R + \alpha_X$, where the standard one, crudely speaking, can be assumed proportional to the kinetic helicity, $\alpha_R \approx -\tau_t \langle \mathbf{u}' \cdot \mathbf{w}' \rangle / 3$. The final balance between the two contributions α_R and α_X determines whether the α

coefficient has the same or the opposite sign to the kinetic helicity. The effect of different terms in the equation (5.85) has been studied in the aforementioned works of [4, 11, 14–16] under some simplifying assumptions, in particular under the neglection of the effects from the cross-response functions G_{ub} and G_{bu}, responsible for the non-equilibrium effects studied here. Assuming that $\alpha = -\tau_t\langle \mathbf{u}' \cdot \mathbf{w}'\rangle/3$ they showed, that the first term $-\alpha\langle \mathbf{B}\rangle \cdot \langle \mathbf{W}\rangle$ always leads to destruction of the cross-helicity. This is no longer true, when $\Upsilon \neq 0$ in non-equilibrium turbulence, since depending on the balance between the $\alpha_{\rm R}$ and $\alpha_{\rm X}$ terms the term $-\alpha\langle \mathbf{B}\rangle \cdot \langle \mathbf{W}\rangle$ in (5.85) may either amplify or destroy the cross-helicity. Furthermore, Yokoi and Hoshino [16] assumed $\beta \sim \langle \mathbf{u}'^2\rangle$ and $\gamma \sim \langle \mathbf{u}' \cdot \mathbf{b}'\rangle$ which allowed them to identify two other terms that always lead to destruction of the cross-helicity, namely

$$-\gamma(\langle \mathbf{W}\rangle + 2\mathbf{\Omega})^2 - \frac{7}{10}\gamma \,\mathrm{Tr}\,(\mathcal{M}^2). \tag{5.86}$$

In addition [16] have described various situations when the terms

$$\beta\langle \mathbf{J}\rangle \cdot \langle \mathbf{W}\rangle, \quad \beta\,\mathrm{Tr}\,(\mathcal{S} \cdot \mathcal{M}), \text{ and } \nabla \cdot [\langle \mathbf{u}'^2 + \mathbf{b}'^2\rangle\langle \mathbf{B}\rangle], \tag{5.87}$$

lead to production of the cross-helicity in the geometry of the tokamak devices. Finally, in the term $-2\alpha\langle \mathbf{B}\rangle \cdot \mathbf{\Omega}$, which in principle may also lead to production of the cross-helicity, we recognize the action of the mean field component parallel to the rotation vector, already identified in (5.71) within the FOSA approach.

The action of all the other terms in (5.85) is difficult to predict and, in general, they can either amplify or destroy the cross-helicity in developed turbulence. The final balance on the right-hand side of (5.85) depends on many dynamical features of turbulence and is expected to be time dependent, however, production of both helicities in turbulence seems likely at least for some periods of time. Indeed, the numerical simulations of [7] demonstrate that the coexistence of the cross- and kinetic helicities is possible in magnetized turbulence in the presence of gravity, density stratification and an imposed magnetic field, $\mathbf{g}\|\nabla\rho\|\mathbf{B}_0\|\mathbf{\Omega}$; the simulations were performed in a periodic box geometry, with the use of the PENCIL CODE (Pencil Code Collaboration). The action of rotation along the direction of stratification has led to generation of kinetic helicity, whereas the action of the imposed magnetic field along the direction of stratification has led to cross-helicity production. Estimates of the $\alpha_{\rm X\,neq}$-effect have also been provided, although in the numerically studied cases the statistical non-stationarity of turbulence was rather weak (see the paper for more details). In conclusion, we can say that in fully developed helical turbulence, that is, in turbulence with strong kinetic helicity, the cross-helicity is rather likely to be produced as well, and at least for some periods of time the two helicities can coexist.

5.4 Summary

The application of the TSDIA method has led to identification of an entirely new contribution to the α-effect in non-equilibrium MHD turbulence, resulting from the infinitesimal-impulse cross responses $\mathbf{u}' \leftrightarrow \mathbf{b}'$, which vanishes in stationary state. The main result is that the non-equilibrium effects in MHD turbulence modify the α-effect by introducing a correction dependent on the square of the non-dimensional cross-helicity $\Upsilon = \langle \mathbf{u}' \cdot \mathbf{b}' \rangle / \sqrt{\langle u'^2 \rangle \langle b'^2 \rangle}$, the kinetic helicity and their history in the MHD turbulence, which takes the form provided in (5.67). This requires coexistence of both the kinetic and cross-helicities in the non-stationary turbulent flow. Such a coexistence is possible and perhaps even ubiquitous in many natural systems, as suggested in section (5.3), where strong cross-helicity production mechanisms identified in earlier works, e.g. [16] have been recalled and discussed, together with the results of numerical simulations of [7]. Finally, we also note that the non-equilibrium effects in turbulence also affect other components of the mean EMF (5.43), that is, the turbulent diffusivity β and the coefficients γ and $\tilde{\gamma}$ in a non-trivial way, through the effect of the Green's cross-response functions G_{ub} and G_{bu}. These interesting effects have not been analysed here, but may be important and perhaps could be investigated in future studies.

References

[1] Yoshizawa A 1985 Statistical theory for magnetohydrodynamic turbulent shear flows *Phys. Fluids* **28** 3313–20

[2] Yoshizawa A 1990 Self-consistent turbulent dynamo modeling of reversed field pinches and planetary magnetic fields *Phys. Fluids* B **2** 1589–600

[3] Yoshizawa A 1998 *Hydrodynamic and Magnetohydrodynamic Turbulent Flows: Modelling and Statistical Theory* (Dordrecht: Springer)

[4] Yokoi N 2013 Cross helicity and related dynamo *Geophys. Astrophys. Fluid Dyn.* **107** 114–84

[5] Yokoi N 2018 Electromotive force in strongly compressible magnetohydrodynamic turbulence *J. Plasma Phys.* **84** 735840501

[6] Yokoi N 2023 Unappreciated cross-helicity effects in plasma physics: anti-diffusion effects in dynamo and momentum transport *Rev. Mod. Plasma Phys.* **7** 33

[7] Mizerski K A, Yokoi N and Brandenburg A 2023 Cross-helicity effect on α-type dynamo in non-equilibrium turbulence *J. Plasma Phys.* **89** 905890412

[8] Kraichnan R H 1959 The structure of isotropic turbulence at very high Reynolds numbers *J. Fluid Mech.* **5** 497–543

[9] Kraichnan R H 1965 Lagrangian-history closure approximation for turbulence *Phys. Fluids* **8** 575–98

[10] Yokoi N 2020 Turbulence, transport and reconnection *Topics in Magnetohydrodynamic Topology, Reconnection and Stability Theory* ed D MacTaggart and A Hillier (Cham: Springer) vol 591 177–265 pp

[11] Yokoi N 2011 Modeling the turbulent cross-helicity evolution: production, dissipation, and transport rates *J. Turb.* **12** N27

[12] Taylor G I 1921 Diffusion by continuous movements *Proc. Lond. Math. Soc.* **20** 196–212

[13] Tennekes H 1979 The exponential Lagrangian correlation function and turbulent diffusion in the inertial subrange *Atmos. Environ.* **13** 1565–7

[14] Yokoi N and Hamba F 2007 An application of the turbulent magnetohydrodynamic residual-energy equation model to the solar wind *Phys. Plasmas* **14** 112904

[15] Yokoi N and Balarac G 2011 Cross-helicity effects and turbulent transport in magnetohydrodynamic flow *J. Phys.: Conf. Ser.* **318** 072039
[16] Yokoi N and Hoshino M 2011 Flow-turbulence interaction in magnetic reconnection *Phys. Plasmas* **18** 111208

IOP Publishing

Non-equilibrium Hydromagnetic Dynamos

Krzysztof A Mizerski

Chapter 6

Potential applications in tokamak research

In fusion reactors such as the tokamak devices the plasma is confined in toroidal geometry by strong magnetic fields. The idea to enhance the magnetic field magnitude by utilization of the dynamo effect has a long history, since such enhancement can lead to stronger confinement and hence stabilization of plasma and increase of its density and temperature, thus making the fusion reactor more efficient; see the reviews by Fisch (1987) [1] and Jarboe (1994) [2]. In this spirit time-dependent helicity injections have been proposed in [3–6] in order to achieve an effective current drive (cf experimental works of [7, 8] and [9] on transient helicity injection and dynamo current drive). Let us briefly review this idea and explain how the non-equilibrium dynamo mechanisms described above could contribute to magnetic field and thus temperature enhancement in tokamaks. The evolution of the total magnetic helicity is governed by equation (2.47) and the helicity density satisfies (cf equations (2.6b) and (2.47))

$$\begin{aligned}\frac{\mathrm{D}}{\mathrm{D}t}\mathbf{A}\cdot\mathbf{B} &= [\mathbf{U}\cdot(\nabla\mathbf{A})]\cdot\mathbf{B}+\mathbf{A}\cdot(\mathbf{B}\cdot\nabla)\mathbf{U}-\nabla\phi\cdot\mathbf{B}+\eta\nabla^2\mathbf{A}\cdot\mathbf{B}+\eta\mathbf{A}\cdot\nabla^2\mathbf{B}\\ &= -2\eta\mathbf{B}\cdot\mathbf{J}+\nabla\cdot[(\mathbf{A}\cdot\mathbf{U}-\phi+\eta\nabla\cdot\mathbf{A})\mathbf{B}+\eta\mathbf{A}\times\mathbf{J}].\end{aligned} \tag{6.1}$$

The first term, namely $-2\eta\mathbf{B}\cdot\mathbf{J}$, accounts for diffusive loss of helicity, however, the flux term involves possible production mechanisms. The idea of transient helicity injection proposed by [4–6] relies on the term

$$\eta\nabla\cdot(\mathbf{A}\times\mathbf{J}), \tag{6.2}$$

and focus has been directed to enhancement of magnetic helicity through utilization of this mechanism, by making the toroidal and poloidal magnetic components oscillatory in time with a phase shift of $\pi/2$. More precisely, in the absence of the mean electric field $\langle\mathbf{E}\rangle_t$ and mean velocity $\langle\mathbf{U}\rangle_t$ the Ohm's law reads (cf (2.5))

$$\mathbf{e}'+\mathbf{u}'\times\langle\mathbf{B}\rangle_t+\mathbf{u}'\times\mathbf{b}'=\eta\mathbf{J}, \tag{6.3}$$

where **J** is the total current field and we have used the temporal mean

$$\langle\psi\rangle_t(t) = \frac{1}{T}\int_{t-T/2}^{t+T/2} \psi(t')\mathrm{d}t'. \tag{6.4}$$

Under the time average equation (6.3) transforms to

$$\langle\mathbf{J}\rangle_t = \frac{1}{\eta}\langle\mathbf{u}' \times \mathbf{b}'\rangle_t, \tag{6.5}$$

This has the status of the mean-field equation and hence the turbulent electromotive force (EMF) must be established by consideration of evolution of the turbulent pulsations. The temporal variations of $\mathbf{u}'$ and $\mathbf{b}'$ are assumed to occur on time-scales mush faster than the diffusive time-scale of the magnetic field, and hence on the turbulent time-scales plasma can be assumed perfectly conducting, which implies

$$\mathbf{e}' + \mathbf{u}' \times \langle\mathbf{B}\rangle_t \approx 0, \tag{6.6}$$

where also the nonlinear term $\mathbf{u}' \times \mathbf{b}'$ is typically neglected (corresponding to the first-order smoothing approximation). Taking a cross product of (6.6) with the large-scale field $\langle\mathbf{B}\rangle_t$ allows us to calculate the turbulent velocity in a tokamak

$$\mathbf{u}' = \frac{\mathbf{e}' \times \langle\mathbf{B}\rangle_t}{\langle\mathbf{B}\rangle_t^2} + \frac{\mathbf{u}' \cdot \langle\mathbf{B}\rangle_t\langle\mathbf{B}\rangle_t}{\langle\mathbf{B}\rangle_t^2}. \tag{6.7}$$

In the approach of [5] the second term is now neglected. This implies

$$\mathbf{u}' = \frac{\mathbf{e}' \times \langle\mathbf{B}\rangle_t}{\langle\mathbf{B}\rangle_t^2}, \tag{6.8}$$

that the turbulent velocity is perpendicular to the mean magnetic field $\langle\mathbf{B}\rangle_t$. Such an approach is then self-consistent, since the term neglected in (6.7) involved the dot product $\mathbf{u}' \cdot \langle\mathbf{B}\rangle_t$ and we may presume such situations to be realizable. Next, introducing the expression for the fluctuational velocity (6.8) into the mean-field equation (6.5) leads to

$$\frac{1}{\eta\langle\mathbf{B}\rangle_t^2}\langle\langle\mathbf{B}\rangle_t\mathbf{e}' \cdot \mathbf{b}' - \mathbf{e}'\mathbf{b}' \cdot \langle\mathbf{B}\rangle_t\rangle_t = \langle\mathbf{J}\rangle_t. \tag{6.9}$$

Taking a dot product with the mean magnetic field and making use of (6.6) which implies $\mathbf{e}' \cdot \langle\mathbf{B}\rangle_t \approx 0$, yields an equation for the current drive along the mean magnetic field

$$\langle\mathbf{J}\rangle_t \cdot \langle\mathbf{B}\rangle_t = \frac{1}{\eta}\langle\mathbf{e}' \cdot \mathbf{b}'\rangle_t, \tag{6.10}$$

The original idea of [5, 6] was to approximate the electric field by the variations of the magnetic vector potential only

$$\mathbf{e}' \approx -\frac{\partial\mathbf{a}'}{\partial t}, \tag{6.11}$$

with neglection of the contribution from the electric potential, introduce single-frequency time oscillations for the poloidal $\mathbf{a}'_{\rm tor}$ and toroidal $\mathbf{a}'_{\rm pol}$ components of the fluctuational magnetic field phase shifted by $\pi/2$

$$\mathbf{a}'_{\rm tor} = \hat{\mathbf{a}}_{\rm tor} \sin \omega t, \qquad \mathbf{a}'_{\rm pol} = \hat{\mathbf{a}}_{\rm pol} \cos \omega t, \tag{6.12}$$

and write

$$\begin{aligned} \langle \mathbf{e}' \cdot \mathbf{b}' \rangle_t &= -\left\langle \frac{\partial \mathbf{a}'}{\partial t} \cdot \nabla \times \mathbf{a}' \right\rangle_t = -\left\langle \frac{\partial \mathbf{a}'_{\rm tor}}{\partial t} \cdot \nabla \times \mathbf{a}'_{\rm pol} + \frac{\partial \mathbf{a}'_{\rm pol}}{\partial t} \cdot \nabla \times \mathbf{a}'_{\rm tor} \right\rangle_t \\ &= \left\langle \frac{\partial \nabla \times \mathbf{a}'_{\rm tor}}{\partial t} \cdot \mathbf{a}'_{\rm pol} + \frac{\partial \nabla \times \mathbf{a}'_{\rm pol}}{\partial t} \cdot \mathbf{a}'_{\rm tor} \right\rangle_t = \left\langle \mathbf{a}' \cdot \nabla \times \frac{\partial \mathbf{a}'}{\partial t} \right\rangle_t . \end{aligned} \tag{6.13}$$

The meaning of the toroidal and poloidal directions in the tokamak geometry is explained in figure 6.1. Consequently, the current drive was estimated as

$$\langle \mathbf{J} \rangle_t \cdot \langle \mathbf{B} \rangle_t \approx \frac{1}{2\eta} \nabla \cdot \left\langle \frac{\partial \mathbf{a}'}{\partial t} \times \mathbf{a}' \right\rangle_t \neq 0, \tag{6.14}$$

which corresponds to the general magnetic helicity production mechanism (6.2) when the currents $\mathbf{J}$ are approximated by $\partial_t \mathbf{A}/\eta$.

We may, however, utilize (6.6) and transform the expression for the current drive in (6.10) back into the more familiar form of (6.5), i.e. $\langle \mathbf{J} \rangle_t \cdot \langle \mathbf{B} \rangle_t = \langle \mathbf{u}' \times \mathbf{b}' \rangle_t \cdot \langle \mathbf{B} \rangle_t / \eta$, where the turbulent EMF is expressed in a standard form corresponding to (2.12). Such form of the mean EMF has been included in the study of dynamo current drive in tokamaks by [10] and in the study of magnetic field relaxation in tokamak-like geometry by [11]. It is now straightforward to see that introduction of the beating effect studied in chapter 3 and chapter 4 should lead to enhancement of the current drive. This could be achieved by transient helicity injection with two close frequencies of oscillation, namely the variations of the magnetic toroidal and poloidal fluxes or

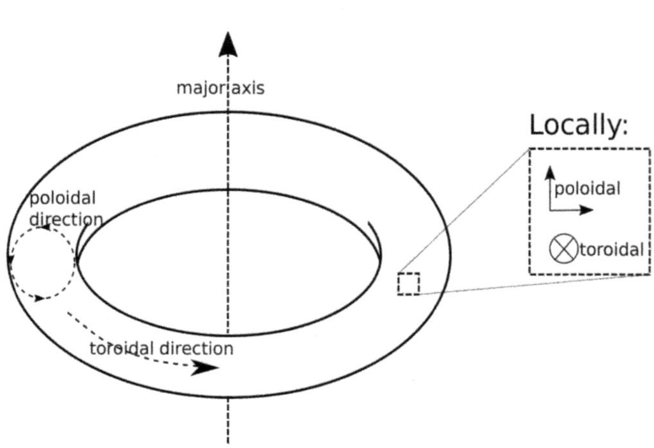

Figure 6.1. Schematic illustration of the toroidal and poloidal directions in tokamak geometry.

equivalently the electric potentials driving the fluxes through Faraday's law of induction should involve two oscillatory beating modes. Indeed, when the fluctuational fields $\mathbf{u}'$ and $\mathbf{b}'$ are composed of two Fourier modes

$$\mathbf{u}' = \frac{1}{2}(\mathbf{u}_1 e^{i\omega_1 t} + \mathbf{u}_2 e^{i\omega_2 t}) + c.\ c., \tag{6.15a}$$

$$\mathbf{b}' = \frac{1}{2}(\mathbf{b}_1 e^{i\omega_1 t} + \mathbf{b}_2 e^{i\omega_2 t}) + c.\ c., \tag{6.15b}$$

with the frequencies satisfying

$$\omega_1, \omega_2 \gg \frac{2\pi}{T} > 0, \qquad 0 < \Delta\omega = \omega_1 - \omega_2 \ll \frac{2\pi}{T}, \tag{6.16}$$

where T is the averaging period (cf (6.4)), the EMF and hence the current drive (cf also (2.65))

$$\begin{aligned}
\langle \mathbf{u}' \times \mathbf{b}' \rangle_t &= \frac{1}{2}\,\mathfrak{Re}\,(\mathbf{u}_1 \times \mathbf{b}_1^* + \mathbf{u}_2 \times \mathbf{b}_2^*) \\
&\quad + \frac{1}{2}\,\mathfrak{Re}\,\langle \mathbf{u}_1 \times \mathbf{b}_2^* e^{i\Delta\omega t} + \mathbf{u}_2 \times \mathbf{b}_1^* e^{-i\Delta\omega t} \rangle_t \\
&\approx \frac{1}{2}\,\mathfrak{Re}\,(\mathbf{u}_1 \times \mathbf{b}_1^* + \mathbf{u}_2 \times \mathbf{b}_2^*) \\
&\quad + \frac{1}{2}\,\mathfrak{Re}\,[\mathbf{u}_1 \times \mathbf{b}_2^* + \mathbf{u}_2 \times \mathbf{b}_1^* + i\Delta\omega t(\mathbf{u}_1 \times \mathbf{b}_2^* - \mathbf{u}_2 \times \mathbf{b}_1^*)],
\end{aligned} \tag{6.17}$$

possess a clear additional contribution resulting from the effect of beat, of the same type as that studied in chapters 3 and 4. Such an approach could provide an effective way of magnetic field and current enhancement, since as shown in chapters 3 and 4 the non-equilibrium mechanism represented here by the second term in (6.17) which can be rearranged into

$$\frac{1}{2}\,\mathfrak{Re}\,(\mathbf{u}_1 \times \mathbf{b}_2^* + \mathbf{u}_2 \times \mathbf{b}_1^*) + \frac{1}{2}\Delta\omega t\,\mathfrak{Im}\,(\mathbf{u}_2 \times \mathbf{b}_1^* - \mathbf{u}_1 \times \mathbf{b}_2^*), \tag{6.18}$$

can be much more efficient than the diffusively controlled one,

$$\frac{1}{2}\,\mathfrak{Re}\,(\mathbf{u}_1 \times \mathbf{b}_1^* + \mathbf{u}_2 \times \mathbf{b}_2^*), \tag{6.19}$$

especially in high conductivity plasma. However, it is important to realize that when the two effects, i.e. the diffusive and non-equilibrium dynamo mechanisms compete, $\bar{\alpha}_{st}\bar{\alpha}_{\Delta\omega} < 0$, one of them needs to dominate over the other in order to get an α-type dynamo and in the case when the standard diffusive dynamo is very efficient $|\, \bar{\alpha}_{st} \,| \gg |\, \bar{\alpha}_{\Delta\omega} \,|$ weaker non-equilibrium effects would only lead to attenuation of dynamo process. Hence tuning these two effects together should lead to the most efficient EMF, which can be achieved by adequate choice of the sign of the phase shift between the two beating modes, $\pm\pi/2$, so that the dynamo mechanisms cooperate to enhance the α-effect (cf chapter 3). Once the large-scale field is enhanced the double oscillatory helicity injection can be switched off and since the field is mostly force-free its decay should be very slow.

The non-stationarity introduced by transient magnetic helicity injection could have additional influence on the turbulent EMF through the $\alpha_{\mathrm{X\,neq}}$-effect based on coexistence of kinetic and cross-helicities in the flow, studied in chapter 5. Plasma in tokamaks should be abundant in kinetic and cross-helicities, as the toroidal components of all three quantities, the flow velocity and vorticity and the magnetic field are typically strong. The fluctuational cross-helicity $\langle \mathbf{u}' \cdot \mathbf{b}' \rangle$ could be injected through any of the two mechanisms proposed by [12] based on the cross-helicity production mechanisms discussed below (5.85). The first one relies on the Reynolds stress and mean magnetic shear coupling which amounts to the coupling of large-scale velocity shear and the gradients of the mean magnetic field (second term in the third line of (5.85)). When both have the same signs positive cross-helicity is produced and negative $\langle \mathbf{u}' \cdot \mathbf{b}' \rangle$ is generated in the opposite case. The cross-helicity would be effectively produced if one of the cases were dominant, thus e.g. if in the regions of largest mean shear the velocity and magnetic gradients were either always of the same sign or always of opposite signs. The second mechanism results from the coupling of the large-scale EMF and the mean vorticity which involves mainly the effect of turbulent diffusivity and coupling of the large-scale currents with the vorticity (the second term in the first line of (5.85)). The toroidal currents are typically very strong in tokamak reactors and the sign of the mean vorticity depends on the direction of the poloidal flow which can also be fairly strong, thus it is very likely that cross-helicity with one specified sign can be produced in such configurations.

References

[1] Fisch N J 1987 Theory of current drive in plasmas *Rev. Mod. Phys.* **59** 175–234

[2] Jarboe T R 1994 Review of spheromak research *Plasma Phys. Control. Fusion* **36** 945–90

[3] Bevir M K and Gray J W 1982 Relaxation, flux consumption and quasi steady state pinches *Proceedings of Reversed Field Pinch Theory Workshop* Report No. LA-8944-C ed H R Lewis and R A Gerwin, Los Alamos Scientific Laboratory pp 176–80

[4] Jensen T H and Chu M S 1984 Current drive and helicity injection *Phys. Fluids* **27** 2881–5

[5] Bellan P M 1984 Physical model of current drive by ac helicity injection *Phys. Fluids* **27** 2191–2

[6] Bellan P M 1985 Mode-beating model of ac helicity injection *Phys. Rev. Lett.* **54** 1381–4

[7] Raman R *et al* 2010 Demonstration of tokamak Ohmic flux saving by transient coaxial helicity injection in the National Spherical Torus Experiment *Phys. Rev. Lett.* **104** 095003

[8] Jarboe T R, Hansen C J, Hossack A C, Marklin G J, Morgan K D, Nelson B A, Sutherland D A and Victor B S 2014 A proof of principle of imposed dynamo current drive: demonstration of sufficient confinement *Fusion Sci. Technol.* **66** 369–84

[9] Victor B S, Jarboe T R, Hansen C J, Akcay C, Morgan K D, Hossack A C and Nelson B A 2014 Sustained spheromaks with ideal $n = 1$ kink stability and pressure confinement *Phys. Plasmas* **21** 082504

[10] Jarboe T R, Victor B S, Nelson B A, Hansen C J, Akcay C, Ennis D A, Hicks N K, Hossack A C, Marklin G J and Smith R J 2012 Imposed-dynamo current drive *Nucl. Fusion* **52** 083017

[11] Moffatt H K and Mizerski K A 2018 Pinch dynamics in a low-β plasma *Fluid Dyn. Res.* **50** 011401

[12] Yokoi N and Hoshino M 2011 Flow-turbulence interaction in magnetic reconnection *Phys. Plasmas* **18** 111208

www.ingramcontent.com/pod-product-compliance
Ingram Content Group UK Ltd.
Pitfield, Milton Keynes, MK11 3LW, UK
UKHW051850100926
8556IPUK00005B/562

9 780750 363587